Project Management
for
Continuous Innovation

Management by Project Mapping

Kern Peng

2018

Pike Publications
Cupertino, California, USA

Published by Pike Publications, Cupertino, California.

Cover image and design: Ilena Peng

While best efforts have been made in preparing this book, the author and publisher cannot assume responsibility for the validity of all materials and the consequences of their use. The strategies, tools and advice contained herein may not be suitable for every situation.

The author and publisher have attempted to trace the copyright holders of all material used in this book and apologize to copyright holders if permission to publish in this form has not been obtained. If any copyright material has not been acknowledged, please let us know so we may rectify in any future reprint.

Author's email for this book: mbpm.innovation@gmail.com
Author's website for this book: mbpminnovation.wordpress.com

Publisher's Cataloging-in-Publication data
Name: Peng, Kern, author.
Title: Project management for continuous innovation: management by project mapping / Kern Peng.
Description: First trade hardcover original edition. | Cupertino, CA: Pike Publications, 2018.
Identifiers: LCCN 2018915200 | ISBN 9781733555708 (hardcover) | ISBN 9781733555715 (epub)
Subjects: LCSH: Project Management. | Technological innovations-Management | Strategic planning. | BISAC: BUSINESS & ECONOMICS / Project Management. | TECHNOLOGY & ENGINEERING / Project Management. | BUSINESS & ECONOMICS / Management Science.
Classification: LCC HD69.P75 P46 2018 (print) | LCC HD69.P75 (ebook) | DDC 658.4/04--dc23
LC record available at https://lccn.loc.gov/2018915200

Printed in the United States of America

First Edition, 2018

This book is dedicated to my wife Patricia,
who teaches me to become the best I can be,

and to my daughters, Ilena and Elena,
who teach me to become a caring person.

Acknowledgments

I gratefully acknowledge all those who helped and supported me in completing this book. Special thanks are due to Mark Cramer and Alvina Nishimoto, who took my course at Stanford and provided valuable feedback and edits. My appreciation also goes to Hal Louchheim, who is in charge of the business course curriculum at the Stanford Continuing Studies. Hal helped me design the course at Stanford which contributes to the organization of this book.

Great appreciation goes to Ilena, my proud daughter, who has spent many hours in proof-reading, offering suggestions and designing the cover of this book. She loves to read and write. Since her sophomore year at Monta Vista High School, she has been an editor of her high school's award-winning El Estoque magazine and later became the Editor-in-Chief in her senior year. She is now studying at George Washington University and also being a staff writer and an editor for the GW newspaper Hatchet. Elena, my other lovely daughter, helped to improve the illustrations of this book. She loves art and drawing. As a sophomore at Monta Vista High School, she is currently the art and web director of La Pluma, her school's literary magazine and the design lead of Res Novae, her school's science magazine.

This book is mainly about corporate innovation and project management. Corporate innovation is defined as a process to achieve breakthrough results for sustainable growth through fulfilling customer and market expectation. It starts with a concept called Management by Project Mapping (MBPM), which reflects the concept of Management by Objective (MBO) from Peter F. Drucker. MBPM is a strategic concept and a systematic methodology that leverage projects to transform an enterprise's operational system, organizational culture, and learning competence to achieve continuous innovation in productive manner. The above-mentioned are well presented under the frameworks namely "Foundation of Sustainable Corporate Innovation" and "Structure of Corporate Innovation".

Dr. Kern Peng, as the author who has been teaching in part-time basis for some postgraduate programs of Stanford University, Institute for China Business of HKU SPACE, and other leading academic institutions in the relevant subject areas of Agile Project Management and Innovation in addition to his full-time leadership role at Intel in Silicon Valley since 2000, highlights that right corporate culture will nurture staff's willingness to go to beyond management expectation to achieve the enterprise's goals, objectives, mission and visions. More importantly, he indicates that the concepts and practices of Agile Project Management for Innovation are much better defined and popular today than they were a decade ago.

Dr. Peng, with his empirical experience of management practice experience in industry, has academically thought through a number of real business cases in today's fast-evolving business environments with international contexts. By using a few examples and/or case studies of Apple, Google, Xiaomi, Tencent and many others, he shares his insights and demonstrates the linkages among project management approach, enterprise's strategic planning, and corporate culture for innovation. In brief, there is a real need for a book of this kind, particularly in the era of digital transformation, collaborative partnership, and open innovation. This book will be widely read and cited.

- Prof. Ning R Liu, Deputy Director, HKU SPACE, Head of Institute for China Business, University of Hong Kong.

Table of Contents

Preface

Chapter 1
Corporate Innovation

Chapter 2
The Project Management Field

Chapter 3
Management by Project Mapping

Chapter 4
Building the Foundations for Innovation

Chapter 5
Project Information Management

Chapter 6
The New WBS for Task Management

Chapter 7
Scheduling for Better Time Management

Chapter 8
Managing Path-finding Projects

Chapter 9
Managing Core Projects

Chapter 10
Managing Continuous Improvement Projects

Chapter 11
Epilogue

About the Author:

Dr. Kern Peng holds doctorate degrees in engineering and business: PhD in Mechanical Engineering specializing in nanocomposite materials, Doctorate of Business Administration in Operations Management and in Management Information Systems. He also holds an MBA in Computer Information Systems and a BS in Industrial Engineering. He has published two books solely in addition to this book and many papers in respected journals and forums such as Engineering Management Journal of IEE, Manufacturing Engineer of IEE, Journal of Advanced Materials, SEMATECH Manufacturing Management Symposium, and Tsinghua Business Review.

Dr. Peng is currently working at Intel, managing the California Validation Center consisting of Santa Clara Validation Center at Intel's headquarter in Silicon Valley and Folsom Validation Center near California state capital Sacramento. He has over 30 years of management experience in engineering and manufacturing and have been a manager at Intel since 1992. He has been accorded more than 100 career awards in the areas of engineering design, software development, excursion resolution, project management and execution, teamwork, leadership and teaching. In addition to regular duties, he serves as a career advisor, mentor for managers, and new employee orientation instructor at Intel.

Since 2000, in addition to working full time at Intel, Dr. Peng has been part-time teaching at least two courses every quarter/semester term at Bay Area universities such as Stanford University, Santa Clara University, University of San Francisco, and San Jose State University. He also travel to Asia several times a year to regularly teach for Hong Kong University. In addition, he has taught courses and lectures for Shanghai Jiao Tong University, Tsinghua University, Zhejiang University, National Taiwan University, UC Berkeley, and San Jose State University.

Project management and innovation have both gained significant attention in recent years, becoming hot topics for management studies and publishing. The Google Ngram charts below show the significant increase in use of both these terms in publishing. Unsurprisingly, when I first offered a course titled "Project Management for Continuous Innovation" in the Spring of 2016 at Stanford University, it was popular enough to gain the highest enrollment in the Continuing Studies program for that quarter.

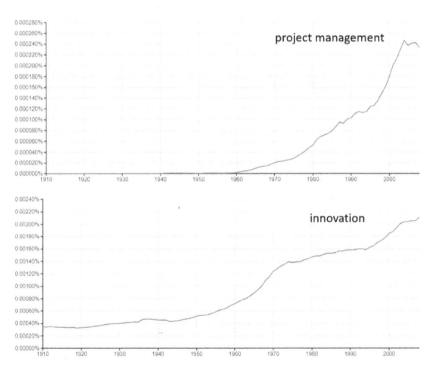

Google Ngram Charts on "project management" & "innovation"

I have been teaching project management every quarter since 2001 as an adjunct at Santa Clara University while working full time at Intel practicing project management. During this time, project management has evolved significantly. The concepts and practices of Agile and Extreme Project Management are much better defined and popular today than they were a decade ago. Teaching the topic now is rather different from the days when Traditional Project Management was mainstream and only a few texts were available. Today, there are many books on project management and equivalently there are many books about innovation, but none have successfully connected both. While designing my course at Stanford, it was apparent that there was a gap to fill.

It is well known that there is a significant correlation between project management and an enterprise's strategic planning; companies make serious efforts to create the right projects and project portfolios for their strategic roadmap. Nevertheless, the effort is both process and result oriented with much of the focus placed on following certain steps to get to the end of projects successfully, and that end result is typically bringing products and/or services to the market. Once the project portfolio is determined, the effort will then transition to managing each project individually and as such, management demands a successful end result for every project.

Project management approaches are being applied at this juncture to ensure the best execution to the company's strategy. Most to the current project management approaches focus solely on techniques and procedures for planning and controlling. These approaches, however, create frameworks that confine project team members to operating within a set of processes and matrices. We should keep in mind that projects are done to ensure the future of the company;

therefore, projects must have innovative results. After all, there is no competitive advantage if a company follows the same projects that others have already done before. Under the current project management frameworks, it is naturally difficult to ask employees to think outside of the box while using a "box" style system to manage projects. Furthermore, because much of the focus of this "box" style system is to achieve success in every project, risk-taking is discouraged because employees may be reluctant to initiate projects unless they feel there is a high probability of success. Inhibited risk-taking constrains learning opportunities and innovation.

Corporate innovation has been a key focus of management studies for the past couple decades, yet a clear definition is still lacking. Many take it at face value and simply believe that innovation means coming up with something new: a new technology, product, business model, marketing strategy and so on. Many reputable media and press organizations, such as the Boston Consulting Group, Fast Company and Forbes, have been publishing "Most Innovative Companies" lists annually for quite some years now. Tellingly, however, there are huge differences between the companies appearing on these lists. In fact, for the year 2016, not a single company appeared in the top 10 of the three lists by BCG, Fast Co. and Forbes. These publishers all have elaborate criteria in measuring corporate innovation but evidently, these measuring criteria vary widely. So if management asks their employees to innovate, some employees might reasonably ask, "According to whose criteria?" If management does not provide a clear direction, employees can't be expected to execute accordingly. Even developing a new technology may not be considered innovative. In fact, the world's top 10 companies in terms of patents granted are seldom in any of the

"Most Innovative Companies" lists. Simply gathering a group of experts to come up with something ingenious that no one has previously done is not necessarily associated with innovation. So what exactly is corporate innovation? The answer is in Chapter One.

One of my favorite books is "The Structure of Scientific Revolutions" by Thomas Kuhn. Dr. Kuhn states that sciences progress through revolutionary paradigm shifts rather than the accumulation of puzzle-solving in an old paradigm. To be considered valid, research studies are typically conducted with the approval of the experts in the paradigm. When a new concept or theory is proposed, the first reaction from the old paradigm is to scrutinize it, often with the intention of suppressing and discrediting it. Certainly, not all new concepts and theories are superior to the old ones, but they are often facing beyond fair skepticism. After all, if someone new to the field develops a better theory, this expert body is facing an identity crisis and will have difficulty maintaining the status quo. It is challenging for any paradigm to willingly accept its fault or incompleteness, and to surrender to a new challenger without a fight. The advancement of science is not only the result of experimentation, problem-solving and accumulation of knowledge, but also under the significant influence of human nature. Similar to scientific development, innovation is also substantially subjected to these same human factors.

I experienced the situation described by Dr. Kuhn when I attempted to publish a book about equipment management in 2011. The mainstream paradigm that guided equipment management was the discipline of maintenance management, which is governed by many associations. When a factory starts up, maintenance is customarily established as a functional department to take care of the equipment. Yet there is a fundamental flaw in the maintenance organizational structure: the departmental and individual objectives of maintenance are not in sync with the objective of the factory. Managers have the desire to grow their departments to progress and survive in the corporate world, yet growing the department means a higher headcount and budget, which must be justified by a higher workload. Thus, growing the maintenance business means more downtime, either in the form of preventive maintenance or repairs. The factory output is negatively impacted by equipment being offline. Conversely, if the manager achieves excellent equipment performance without any downtime, the maintenance department may be significantly reduced or cease to exist.

It is a dilemma that can only be addressed by moving away from the maintenance organizational structure. In addition, one of the main objectives of maintenance management – extending the life of equipment – is somewhat obsolete. It is a fact that in many high-tech companies, equipment replacement rarely occurs because of the end of the equipment's natural life. This new phenomenon is also seen in the general consumer market where most personal electronic devices, such as cell phones and computers, are replaced not because they are broken, but because a better version has arrived on the market.

I proposed that traditional maintenance management principles were no longer effective in the fast changing business environment when I submitted my book proposal to the largest publisher in maintenance management. It caused quite a bit of disturbance and they rejected my proposal for "not wanting to alienate their reader base." The Editor-in-Chief was on my side and wished that his organization "wasn't so conservative," so he recommended other publishers that were outside the maintenance field and my book was finally published by Productivity Press. Developing a new idea is exciting, but the process of getting acceptance is often accompanied with rejections from established experts. Innovation requires persistence.

This book is about corporate innovation, which is the ultimate focus of project management. The purpose of this book is to 1) further define corporate innovation beyond the conventional meanings, 2) provide a system approach model that strategically utilizes project management to achieve continuous innovation, and 3) apply different project management approaches to different types of projects based on corporate strategy.

Most project management books are targeted toward project managers and practitioners. This book starts with a concept called Management by Project Mapping (MBPM), which mirrors the concept of Management by Objectives (MBO). At the strategic level for the C level audience, MBPM is a systematic approach that utilizes projects to transform an organization's system, culture and learning capability to consistently achieve continuous innovation. Strategy without execution, however, is merely a wish, so this book will also cover the proven tactical approaches that are suitable for different types of projects derived from project mapping at the strategic level. The target audience of this book includes all levels of management, strategy managers and people managers as well as task execution managers.

Different audiences desire different levels of details; therefore, this book is designed to help different audiences get to what they want to read. It presents the content in three levels of details with the special page number designations, T#, M#, and L#, shown on the top of each page in addition to normal page numbers on the bottom. Each chapter consists of all three sections. The strategic level audience may not have time to read through hundreds of pages to get the key points buried throughout the book. As such, they can simply follow the "T" section, which is at the strategic level with main ideas presented (answering the "what?"). The "M" section will elaborate the concepts in further detail and provide reasons and justifications for the concepts (answering the "why?"). The "L" section provides guidance for practicing the concept with tips and tricks (answering the "how?").

Printed books often represent a one-way information flow. A book in the hands of a reader is a final text with fixed information that does not allow readers to gain additional knowledge or information beyond the printed content. This book, however, offers a unique feature. At the end of each chapter, there is a "Q" section with critical questions on the topic. It may look like the end of chapter section with exercise questions in a typical textbook, but it is different as the questions in the "Q" section are typically beyond the content of the chapter and obvious answers cannot be found in the text. This is a method of inviting you to think deeper and you are welcome to share your thoughts on the questions by email MBPM.Innovation@gmail.com. For many of the questions, I don't have definite answers but will share my up-to-date thoughts and answers with you. My responses will be updated as I am continuing to learn and practice in the world of management. Evidently, doing so is quite a significant commitment on my part so I need to declare some requirements to ensure the feasibility of achieving this promise:

1) One question at a time and state the Question # in the subject line of the email. This allows me to sort the emails quickly.

2) Provide a scanned copy of the book purchase receipt the first time you use an email to send in a question. After all, if you did not purchase the book, I have no obligation to respond.

3) My response will be sent to you between 2-4 weeks after I receive your question. This is because this book may be used as a textbook and instructors like myself may assign the questions as homework. I would like to avoid the chance of students turning in my responses to their instructors instead of developing their own answers. As long as instructors set the homework due date within a couple weeks of the assignment, students should not have gotten the answers from me.

Such unique organization offers readers the opportunity to choose various methods of obtaining knowledge from this book, giving choices for readers with different learning styles. Readers can follow the traditional path of reading chapter by the chapter or choose to read all the "T" sections first, the "M" sections next and the "L" sections last. Other readers may get high level directions by reading the "T" sections then going directly to the "L" sections for practical executions, or may even just go the "Q" sections and interact with me through email. I hope this unique approach will further enhance your enjoyment from this book. This book is about innovation and I am experimenting with a new way of presenting the materials. Whether or not this approach is innovative is determined by you. No creator can declare their own creation an innovation. Only the users can. Innovation is defined by users.

Project Management
for
Continuous Innovation

Management by Project Mapping

Corporate Innovation

Defining Corporate Innovation

The world is changing and the business world must changing along with it.

Corporation must innovate to keep up.

"Innovation is everything..." by Robert Noyce

The Main Entrance of Intel's Headquarter Building

Innovation is not just simply means inventing something new.

There are many ways to measure corporate innovation. Go to Chapter 1 M section (M1) for details about measuring corporate innovation.

It is rare and difficult to consistently achieve innovative results. Go to Chapter 1 L section (L1) for details about the challenging journey of innovation.

What is corporate innovation?

Innovation is achieving breakthrough results!

... but breaking through what?

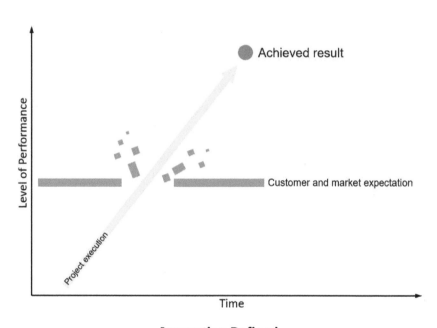

Innovation Defined

The answer is:

customer and market *expectation*.

A company would not be considered innovative if it only meets the customer and the market's expectations. It is the "wow" factor and the surprise elements that contribute to innovativeness in the eyes of the consumers.

It is important to understand customers' expectations will not stay the same... the more you innovate, the more the customers expect from you and thus, the more difficult it is to be innovative. Continually exceeding the expectation of the customers and the market is key.

Effectively mapping and managing projects will enable a corporation to achieve continuing innovation. Go to Chapter 2 T section (T2) to understand the importance of the relationship between project management and innovation.

Corporate Innovation

Measuring Corporation Innovation

Could you name some of the innovative companies?

Perhaps:

> Apple,
>
> Google,
>
> Tesla...

Why do you consider these companies innovative?

Perhaps:

> New products,
>
> New markets,
>
> New way of doing business...

What criteria are you using?

Perhaps:

> Creating values,
>
> Making money,
>
> Changing our lives...

Now let's look at how some of the most reputable organizations and media measure corporate innovation. Boston Consulting Group, Fast Company and Forbes all publish an annual list of the world's most innovative companies [1] [2] [3].

2016 World's Most Innovative Companies **Boston Consulting Group**		2016 World's Most Innovative Companies **Fast Company**		2016 World's Most Innovative Companies **Forbes**	
1	Apple	1	BuzzFeed	1	Tesla Motors
2	Google	2	Facebook	2	Salesforce.com
3	Tesla Motors	3	CVS Health	3	Regeneron Pharmaceuticals
4	Microsoft	4	Uber	4	Incyte
5	Amazon	5	Netflix	5	Alexion Pharmaceuticals
6	Netflix	6	Amazon	6	Under Armour
7	Samsung	7	Apple	7	Monster Beverage
8	Toyota	8	Alphabet	8	Unilever Indonesia
9	Facebook	9	Black Lives Matter	9	Vertex Pharmaceuticals
10	IBM	10	Taco Bell	10	Biomarin Pharmaceutical
11	Bayer	11	Robinhood	11	Amazon
12	Southwest Airlines	12	Universal Studios	12	Arm Holdings
13	Hewlett-Packard	13	Huawei	13	Naver
14	BMW	14	Cyanogen	14	Fleetcor Technologies
15	General Electric	15	InMobi	15	Netflix
16	Daimler	16	Novocure	16	Shanghai RAAS Blood Products
17	Uber	17	Bristol-Myers Squibb	17	Rakuten
18	Dupont	18	Amgen	18	Asian Paints
19	Dow Chemical	19	Spotify	19	LG Household & Health Care
20	BASF	20	General Electric	20	Verisk Analytics
21	Airbnb	21	Warby Parker	21	Amorepacific
22	Under Armour	22	Riot Games	22	Coloplast
23	Gilead Sciences	23	Farfetch	23	Marriott International
24	Regeneron Pharmaceuticals	24	Everlane	24	Illumina
25	Cisco Systems	25	Kit and Ace	25	Red Hat

The World's Most Innovative Companies Lists 2016
Sources: The Boston Consulting Group, Fast Company and Forbes

Remarkably, however, only 2 companies appeared on the top 25 spots of all three lists. If we only consider the top 10 companies, then not even one company shares all three lists!

Why do these organizations measure innovative so differently?

These organizations of course all use comprehensive criteria to generate the lists as objectively as possible and they can certainly defend what they used with reasonable logic.

Why does the difference in measuring corporate innovation matter?

If you ask your employees to innovate, they might ask you a question in return -

"Which Innovative Companies list are we targeting?"

How would you respond? Telling employees to just do it and hopefully we will get onto one of the lists? Innovation should not be just a word in a slogan. If you are not clear, how can you ask employees to execute? Without a specific direction, you cannot expect your employee to achieve the desired results.

From the business perspective, a company needs to make profit to survive. For those who focus on the financial results, corporate innovation should reflect on the bottom line business performance of the company. The profitability of the company, however, is obviously not a key criteria for measuring innovativeness, at least not on those three lists. Tesla Motors, the only company placed within the top 3 in two of the three lists, has yet to make a profit since it was founded. On the Fast Company list, where Telsa Motors was not in the top 3, BuzzFeed took the first place, which is clearly not due to profitability either.

From a technical perspective, the word innovation is probably associated the most with high-tech industries. For those who focus on R&D and technical results, corporate innovation could be measured by the number of patents granted. However, innovation is not just simply gathering a group of technical experts and developing something that no one has done before. Many technically genius ideas do not make it to the market.

Rank	Patent Granted	Company	Country
1	36264	International Business Machines Corp	United States
2	25299	Samsung Electronics Co Ltd	Korea
3	18853	Canon KK	Japan
4	13990	Microsoft Technology Licensing LLC	United States
5	12456	Sony Corp	Japan
6	12052	Toshiba Corp	Japan
7	11782	Qualcomm Inc	United States
8	11238	Google Inc	United States
9	10363	LG Electronics Inc	Korea
10	10339	Panasonic Intellectual Property Management Co Ltd	Japan
11	9155	Intel Corp	United States
12	8954	Apple Inc	United States
13	8654	General Electric Co	United States
14	8196	Fujitsu Ltd	Japan
15	7884	Seiko Epson Corp	Japan
16	7715	Hon Hai Precision Industry Co Ltd	Taiwan
17	7553	Ricoh Co Ltd	Japan
18	7113	Taiwan Semiconductor Manufacturing Co (TSMC) Ltd	Taiwan
19	7069	Toyota Motor Corp	Japan
20	6916	GM Global Technology Operations LLC	United States
21	6642	Samsung Display Co Ltd	Korea
22	6488	Telefonaktiebolaget L M Ericsson	Sweden
23	6210	Hewlett Packard Development Co LP	United States
24	5353	Honda Motor Co Ltd	Japan
25	5345	Broadcom Corp	United States

The World's Top 25 Patent Generating Companies 2012-2016
Source: IFI CLAIMS [4]

Let's examine the patents granted by the United States Patent and Trademark Office (USPTO) and the list of companies with the most patents granted. Since the patent application and granting process requires a rather long period of time, a five-year period is used and the top 25 patent generating companies from 2012 to 2016 are listed [4]. It is quite different from the other three lists shown earlier. With the addition of this new list, there is not even a single company that appears on all four lists. Why are most of these technically aggressive companies, investing huge amount of money in R&D and demonstrating excellent results in patent generations, failing to be considered innovative?

If corporate innovation isn't measured by profitability, or technical capability, then how is it measured?

It is simply the perception of the customers and the market. Companies that are considered innovative must have a great focus on their customers and the markets. They understand the needs of their customers and markets and, more importantly, not just meeting those needs but exceeding the expectations of their customers.

Innovation requires a lot of effort. It is not an easy journey, especially if a company is aiming at maintaining innovativeness and building long lasting enterprise. Understanding the journey itself will gain more in-depth knowledge about innovation.

Continue reading the next section to find out more about the journey of innovation and how to plan for it or go to Chapter 2 for the importance of project management and its relationship to innovation.

Corporate Innovation

The Journey of Innovation

Innovation means achieving breakthrough results and more specifically breaking through customer and market expectations, which is difficult to maintain constantly. The process of achieving these results continuously is defined as the journey of innovation.

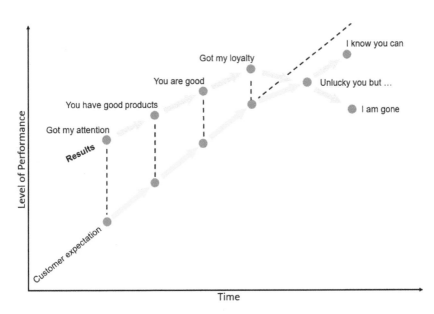

The Journey of Innovation

As shown in the above figure, the first innovation gets the customer's attention and several consistent innovations are then needed to achieve customer loyalty. Customers also expect more as a company innovates so the bar is rising for the next generation of products or services offered. Simply coming up with a product that is better than the previous generation will not be considered innovative if it just meets the customer expectations. Engineers may feel that a new product incorporates many technical breakthroughs and exceeds management's expectation as well. However, exceeding the company's own expectation does not count: only exceeding the expectations of the customers and market is considered innovative.

Using the analogy of dating that leads to engagement and marriage, one must act gradually to exceed the expectation of the targeted significant other in order to make progress in the relationship. For instance, if a guy gives a diamond ring to a girl on their first date, no small gifts in the second date will be sufficient to meet the high expectation set at the first date. Many startup companies went all-in to attract their customers and a few succeeded. However, many of these companies failed because of lacking follow-on products that were needed to continuously exceed the customers' expectations that they initially set. It is a well-known fact that most startups fail within their first year.

For an entrepreneur who wants to establish a long-lasting company, having only one good product is not enough. The company needs a strategic plan with follow-on products that gradually exceed its customer expectations. This is like people pursuing a long-lasting relationship that leads to marriage. Of course, there are a few entrepreneurs who intentionally go all-in on one product, aiming at exceeding customers' expectations by far to get a lot of good press and attention, and then sell the company at the peak of good publicity and move on to something else with the money. However, most entrepreneurs lack clear objectives and long-term plans when starting their companies and take their chances with fate to do the best as they go. Using Xiaomi in China as an example, Xiaomi's cellphone market growth in China was phenomenal for the first couple years, but was then followed by negative press. Its CEO, Lei, publicly stated that Xiaomi was doing so well in the beginning that it set an unrealistic high expectation to meet continuously. He was asking its customers to be patient and saying that the company had been back to normal pace as it should be. In other words, he was attempting to reset customers' expectations.

Customer expectations can shift and will give the company some slack for a limited time if that the company has already gained their loyalty. As such, a one-time miss may not result in damage to the company's image and result the loss of customers. Of course, that also depends on the size of the gap between the expectation and the actual reality. Back to the analogy of dating, if a couple's relationship has progressed to the point of getting engaged or married, one small mistake would not result a breakup. However, company and customers relationships are typically not as strong as marriages, so a couple misses are likely to result in losing customers. No doubt, a few companies have done very well in achieving strong customer relationships close to marriage. However, in any case, failing to meet expectations for the first time is an important inflection point for a company, as the company must now try to regain the customers' trust or it could end up losing them. It also means a reset in customer and market expectation, which may not necessarily be a bad thing.

There are only limited examples of companies that are able to keep innovating continuously. Apple Inc. is a good example but even it has ups and downs in the innovation journey. Since shifting their focus into the consumer electronics market with the iPod/iTunes, iPhone, and then the iPad, it has been considered one of the most innovative companies in the world. However, its future journey is filled with uncertainties and in my opinion, it is at the inflection point. We are waiting for the next "wow" product from Apple. Innovation is determined by the market and consumers. Exceeding the company's own goals in achieving its vision may not be considered innovative. It is difficult to keep up with the innovation journey as companies must continue to thrust the effort to exceed market and customer expectations. The crucial question is thus "how can a company keep innovating time and time again?"

Before answering this question, let's look at the two main paths toward innovation. The first is leader driven innovation, which is often seen in startup companies. The founder has a novel idea and puts it into action to become an entrepreneur. If the founder's idea is not accepted by the market, the company fails and the innovation is not recognized. If the company attracts customers and is noticed by the market, it must have had a unique product or service that is considered innovative. That is the starting point of the innovation journey. Typically, a company cannot rely on a single person - the founder in this case - to keep innovating. The vision needs to be in constant renewal to keep exceeding customer and market expectations.

After founding the company, the founder must divert efforts toward the logistics of running the company and can no longer focus extensively on pursuing innovative ideas. One way to cope with this problem is to find a professional executive to run the company while the founder continue to focus on product development and innovation. Steve Jobs tried that by convincing John Sculley, then CEO of Pepsi, to run Apple with his famous line: "Do you want to sell sugar water for the rest of your life? Or do you want to come with me and change the world?" The results were not what Jobs was hoping for as he ended up being pushed out of Apple. Bringing someone from outside a company can often create a power struggle between the new executive and the existing management team.

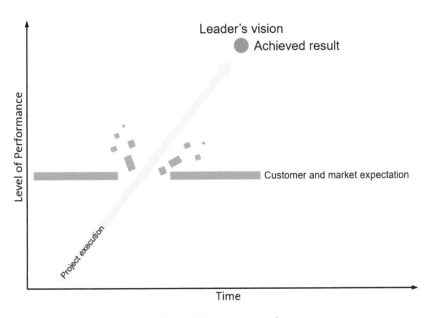

Leader Driven Innovation

In most cases, as a company grows, the founder tends to hire people who are good at execution because the founder requires individuals who can execute his or her vision. What then happens if the leader steps down? Typically, those who demonstrated success in their jobs would take on leadership positions. The company would then be led by people who are good at execution but lacking vision. These new leaders are efficient if they know what to do, but in the absence of a clear vision they would take a typical textbook "user-centric and customer orientated" approach to finding out what customers want and then design products or services based on the users' needs. Only doing what users tell you offers no element of surprise for the users – and an element of surprise is critical for being considered innovative. Therefore, this path of innovation can seldom be sustained.

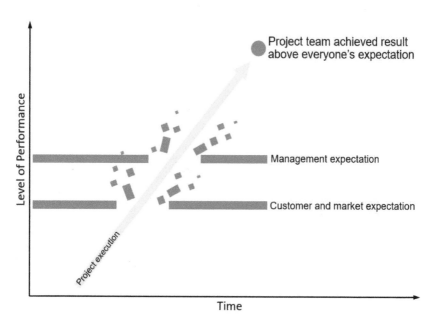

Corporate Culture Driven Innovation

Companies that cannot maintain innovation will remain small or eventually be out of business. Companies that continue to innovate have more potential to grow into large corporations, but no large corporation can sustain its innovativeness by depending on a single individual. The desired approach to sustaining long-term innovation is a corporate culture driven innovation. As demonstrated in the figure, management has a vision that exceeds the customer and market expectations, and more importantly, the project team achieves the performance result above all expectations including the expectation from management. Google and Tencent are two successful examples in using this approach to sustain innovation.

The right corporate culture attracts devoted talents and creates an environment for a learning organization, which creates sustainable capability. It is the culture that fertilizes employees' willingness to go above and beyond management expectation to achieve innovative results. People who seek achievement will have the desire to learn and eventually gain the ability to do so. With committed and talented employees, innovation will come. Innovation, to every single employee in a company, simply means producing a surprising result that exceeds the expectation of customers who may be external or internal.

If a company wants to achieve consistent innovative results from projects, the project management approach cannot just focus on meeting the targeted results of a few projects. Instead, it should take the systems approach to utilizing project management to build the foundation for innovation. The next chapter will discuss project management and its relationship with innovation.

Chapter 1

Corporate Innovation

Exercise Questions

E-mail your thoughts on the questions to MBPM.Innovation@gmail.com I will share my thoughts and answers to the questions.

Rules: (For details and reasons, please read Preface Page xvii)

1) One question at a time and state the Question # in the subject line of the email.

2) Provide a scanned copy of the book purchase receipt the first time you use an email to send in a question. This won't be necessary for future questions using the same email.

3) My response will be sent to you between 2-4 weeks after I receive your email.

Q1-1. Should you implement a program to drive innovation at your company? If so, are you planning to do it as a short-term initiative or as a long-term endeavor?

Q1-2. If you decide to push innovation at your company, what direction should you set and how do you measure the success?

Q1-3. In your opinion, which media provides the best ranking for the world's most innovative companies?

Q1-4. Should technical companies push patent generation and using rewards as part their innovation program in promoting innovation? If so, should the quantity of applications or acceptances be used to measure the success of the program?

Q1-5. Could you find some examples of leader-led innovations? Do you think they are successful?

Q1-6. Could you find examples of corporate culture driven innovations? Do these companies setup their organization in such way at the beginning or do they transition from the leader-led innovation path? Do they have an effective organization setup for promoting innovation?

Note to Readers

Just a reminder on the organization of this book, which is described in greater details in the Preface page xvi to page xviii. I encourage you to read the section in case you missed it.

In each chapter, the T section generally describes the concept at a high level – the what. The M section explains the reasons behind the concept – the why. The L section provides the guidance on practicing the concept – the how.

The Q section invites you, the reader, to think beyond the writing of the chapter. Many of the questions came to me while writing the book and I didn't have definite answers at the time, but I will share my up-to-date answer with you within 2-4 weeks after I receive your email. I invite you to share your answer with me. Instruction for contacting me is documented on the Q sections of all chapters.

Chapter 2

The Project Management Field

Project Management and Innovation

Organizational work can be categorized as projects or operations.

Projects are temporary endeavors undertaken with the purpose of achieving a unique end-result, which is the process of going from 0 to 1; while operations are ongoing, repetitive endeavors that produce many identical, or nearly identical end items, which is the process of going from 1 to n.

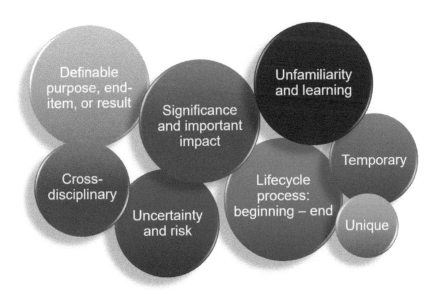

Key Project Characteristics

In a typical company (the exception being companies that get paid for doing projects for others, such as IDEO), people who are working on projects are considered overhead since their salaries are paid by profits made from operations.

Why then are companies spending their resources on projects, which do not directly generate revenue?

The simple answer is that it is for the future of the company.
As such, it is crucial to have innovative projects. If the results of the projects are not innovative, the company will not gain competitive advantages in the market, thus jeopardizing its future.

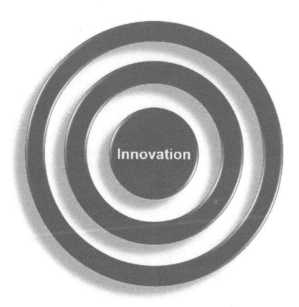

The Target of Project Management

The project management field is becoming more and more popular as companies realize that projects, as strategic endeavors, deserve a significant investment of both time and effort to ensure proper execution.

Most of the current project management approaches focus on the success of individual projects and lack a built-in mechanism for promoting corporate innovation. On the contrary, these approaches often impede innovation with their well-structured processes.

Read the next section (M2) of this chapter to understand the reasons or go to Chapter 3 to find out a new concept called "Management by Project Mapping (MBPM)," which can systematically and strategically utilize projects to achieve continuous corporate innovation.

The Project Management Field

Strategic Use of Project Management

Projects are the building blocks in the design and execution of strategies for an enterprise to achieve its vision. And that vision must include innovation.

STRATEGIES   PROJECTS   VISION

Projects as the Link between Strategic Planning and Vision

It is well known that there is a significant relationship between projects and an enterprise's strategic planning. Companies devote serious efforts to developing their project portfolios and planning their project roadmaps. Much of the efforts, however, are too focused on tactical level approaches and applications. For example, the project selection process focuses on picking the right projects that meet the needs of the market or customers. When designing a project portfolio, considerations are often based on customer requirements, complementary to product lifecycles, factory capacity, or the needs for different market segments. The approaches are process oriented, with an overwhelming emphasis on results but tending to be very inconsistent in achieving innovative results. There are many examples of companies that had a few innovative products and services but failed to keep up with the markets. As we have seen in the journey of innovation, being successful in a few projects and doing well with a few products is not enough to sustain a company's long-term competitiveness.

Most of the current project management approaches aim to achieve planned results for a given project with a main focus on techniques and procedures for planning and controlling. These approaches create a framework that confines project team members to operate within a set of processes and measures: set a scope, develop a charter document, have a plan, control changes, report status and so on. When things are not going well, more processes are then added to remedy the situation. This may be one of the contributing factors for the growth of project management. The number of processes in PMI's PMBOK guide has grown from 37 processes in its first edition to 49 processes in the latest edition (6th Edition published September 2017) [5]. Many project management practices are bureaucratic and confine the project team members to operate within certain perimeters.

As shown in the figure on the next page, these practices create a control system, like putting guardrails to ensure that the execution of the project is within certain limits. Management is also making sure that these limits are well-known to the project team members by rewarding them when they are doing well and taking corrective actions when they are doing poorly.

It is difficult to ask project team members to think outside of the box while using this "box" style system to manage projects. Policies and procedures like this can only create a compliance-based system. If we want to create a value-based system in which employees are motivated to perform beyond management expectations, different project management approaches must be used.

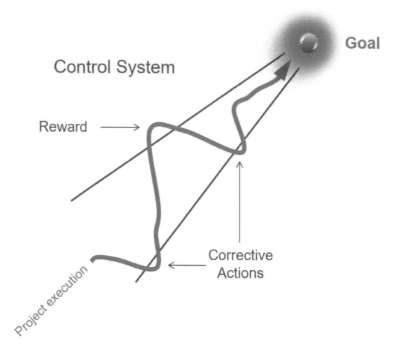

Project Execution under PM Control System

In addition, much of the focus of such a system is to achieve a successful result in every project, which is unrealistic and often unnecessary. Innovation requires a great amount of effort in exploring and experiencing. A research study done by Stevens and Burley demonstrated that it takes about 3,000 raw ideas to get one commercial success [6]. The illustration on the next page summarizes their research findings and shows that 3,000 ideas lead to 125 projects. Many of these projects did not materialize, but they certainly created knowledge serving as the foundation for the success. Current project management focuses on applying processes and techniques to achieve successful results in the given projects. Management applies project management tools with the objective of achieving a high project success rate. They missed the fact that not all projects need to be done in a "correct" way. More importantly, not all projects need to reach a successful end. In fact, most of the projects don't, as the ratio of 125 to 1 demonstrated by Stevens and Burley's research.

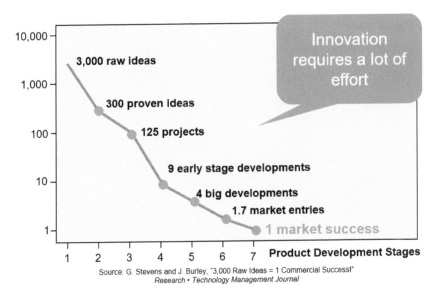

From Idea to Market Success

Over emphasis on project success discourages risk-taking because employees may not want to initiate projects unless they feel the chance of success is high. Inhibiting risk-taking constrains learning opportunities and innovation. In other words, failures should be accepted on the path to innovation. It is okay to fail in some projects, but we also want success in certain projects too. Therefore, the first step in project management should not be selecting and initiating projects that have higher success rate, as current project management approaches do. It should be identifying and categorizing projects so that we know which ones are okay to fail and which ones must be successful.

Go to Chapter 3 to understand the connotation of MBPM, which is designed with this exact purpose in mind. Obviously, one single project management approach cannot be used for projects with totally different desired outcomes. Therefore, we need to understand the different project management approaches currently being used. It sets the foundation for applying the right approaches to the right category of projects. Read the next section (L2) of this chapter for an overview of Traditional, Agile and Extreme Project Management.

The Project Management Field

Project Management Approaches:
Traditional, Agile and Extreme

The practices of modern project management started in the 1900s and were generally applied in civil engineering projects. The first major tool widely applied in project management was the Gantt Chart, which was developed by Henry Gantt in the 1910s. In the 1950s, project management was formally recognized as a distinct discipline, but it was mostly used in government related activities in defense, space exploration and civil construction.

Many widely used project management tools, such as Work Breakdown Structure (WBS), Critical Path Method (CPM) and Program Evaluation and review Techniques (PERT), were developed from the 1940s to 1960s through joint partnerships between US government agencies like the Department of Defense and Navy, and private companies such as DuPont and Lockheed. In 1969, the Project Management Institute (PMI) was formed and has become the most recognizable project management professional organization today. PMI publishes "A Guide to the Project Management Body of Knowledge" as a standard to apply to the management of most projects.

It wasn't until the 1980s that private industries started to widely apply project management practices. Partly due to fierce competitions from Japan and the Asia Dragons, US companies were widely restructuring and started the movement of leveling the organization structure. Many managers found their new jobs in project management as a result. Project management gained the needed resources to develop in a more organized fashion.

Since then, as evidenced by its presence in publications for about two decades, project management experienced fast growth in practitioners until early 2003 when it started to level off. This can be seen in the Google Ngram chart below which shows the occurrence of the phase "project management" in publications. During this time, many project management techniques were refined and new approaches, such as Extreme Project Management and Agile Project Management, were developed and widely practiced in addition to Traditional Project Management.

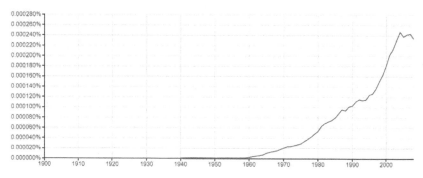

Google Ngram Chart on Project Management

General project management methods are currently categorized into three major approaches: Traditional, Agile and Extreme [7]. Robert Wysocki does a great job presenting and comparing the three approaches in his book *Effective Project Management: Traditional, Agile, Extreme*, now in its 7th edition. For those who wish to understand the details of these approaches, reading Wysocki's book is highly recommended. We will not go into the details here but rather provide an overview and the pros and cons of these project management approaches.

Traditional Project Management

Traditional Project Management is essentially based on the notion of planning the work and then working the plan. Learning the Traditional Project Management means learning the tools and techniques to manage the project lifecycle and processes. At approximately the same time that project management was getting popular, the systems approach movement was occurring as well. Early project management professionals saw the similarity of both and considered projects as complex systems, so they used the system development cycle to build the project life cycle model. Early project management textbooks generally had chapters to cover systems thinking, systems approach, systems development, systems engineering, etc. The typical project life cycle model has four phases: conception, definition, execution and implementation. PMI developed its own process model to manage the project from the beginning to the end with five phases: initiating, planning, executing, controlling and closing.

The project life cycle, also regarded as the systems development cycle, determines the end of each phase by the completion of one or a set of deliverables. The conception phase is done when the project proposal is developed and presented. The definition phase is completed as the project plan is done. The execution phase is finished when the defined system is built, whether it is a product or a service design. The implementation phase is done when the system is successfully handed over to the users by finishing documentation, training and so on.

The project life cycle model has been widely used and expanded for managing projects in specific fields, such as software development life cycle, new product introduction life cycle, construction project life cycle, etc. These live cycle models provide general understanding, languages or even standards for the professionals in their particular fields to work together efficiently. For instance, the software development cycle includes milestones like alpha and beta testing. Any software project participant knows the meaning of these terms and can understand the project's exact status. Similarly, in construction projects, construction professionals use typical milestones like 30%, 60%, 90% and 100% design reviews for managing the design progress.

The primary criticism of the project life cycle model is how its phases are based on deliverables but not processes. In other words, the model needs to be executed in a sequential order; one phase starts only when the preceding one finishes. PMI views project management as a series of processes spread across five phases as shown in the figure on the next page. A process is defined as a series of actions that apply tools and techniques to inputs to produce outputs. PMI's PMBOK is organized based on this process model that consists of 49 processes in the latest 6th edition [5].

Whether deliverable-focused or process-oriented, Traditional Project Management emphasizes the importance of planning and controlling, and especially controlling. The difference between the phases of the project life cycle model and that of the process model is the additional "controlling" phase, which covers nearly the entire length of the project timeline. As mentioned in the previous chapter, the controlling aspect of management tactics confines innovation, which makes it less ideal in many situations. Traditional Project Management, however, does provide a structural approach for complex systems so it is still widely practiced in many industries and companies, especially in government related projects such as those in civil, infrastructure and defense areas where control is deemed important.

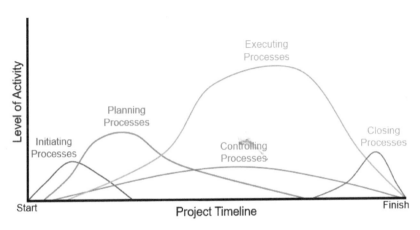

The Project Process Model

In such situations where Traditional Project Management approach is required by management, projects would not get funding and approval without completing a thorough proposal and a comprehensive plan. Therefore, extensive efforts are put into the initiating and planning phases, which typically take quite a long period of time. However, it is the doing phase that actually creates the bottom-line value - initiating and planning are essentially administrative overhead activities. There is no disagreement that planning makes doing more effective, but to what extend are such administrative efforts needed? Once such a comprehensive plan is done, what is the probability that something

31

changes and the plan needs modification? It is extremely high, and the longer the project, the higher the likelihood of changes. That is why most large civil projects have multiple plan revisions and such projects are typically over budget and behind schedule.

The typical resolution for these problems consists of putting more measures and processes in place to ensure plan accuracy and tighter control to mitigate risks. It is like adding more administrative effort to address administrative inefficiencies, and fighting bureaucracy with bureaucracy. It is no surprise that the field of project management is growing with more processes and techniques to address the problems that are coming back again and again. If the fundamental management approach does not change, we can only hope that an experienced project manager and team, along with luck, all come together well enough to save the day.

Traditional Project Management is still useful in some situations but it must be executed in a simpler manner. In later chapters of this book, Traditional Project Management tactics are applied in many circumstances. The plan the work and work the plan mentality is very much alive and applicable. However, it should be less process oriented and more people oriented with the intent of reducing administrative efforts.

Extreme Project Management

Let's now discuss Extreme Project Management before we discuss Agile Project Management, since the Extreme Project Management method was proposed before Agile and many people consider Agile a hybrid of

Traditional and Extreme Project Management. Both Extreme and Agile approaches evolved from the fast growing and constantly changing software industry.

Extreme Project Management takes a flexible, open and non-deterministic approach to managing projects. It focuses more on the human side of project management than the techniques and processes. It started in the late 1990s with the dot-com and internet boom when time to market was considered a key competitive advantage. As a result, many companies skipped the initiating and planning phases and used the "just do it" test-first approach. It encourages trying and exploring without many defined processes or steps. Some project management professionals attempted to define Extreme Project Management by coming up with models such as INSPIRE (which stands for INitiate, SPeculate, Incubate, and REview), but these models are vague, and in my opinion, such attempts defect the purpose of the Extreme concept by using traditional management approach to put a frame around the concept. Once the model is defined with specific steps, the open and non-deterministic characteristics vanish.

In reality, the practice of Extreme Project Management is rare, as management is typically very uncomfortable with such a free-style approach with unknown budget and schedule. After all, most companies have to report their performance to stockholders or investors on a regular basis. The dot-com bust, with its gross project failure rate, also put a stop on many companies that practiced the Extreme approach. Nevertheless, it forced the Extreme Project Management practitioners to think of better ways to manage projects without going back to the very structured Traditional Project

Management approach. As the result, Agile Project Management approach surfaced and it is becoming more and more popular.

Should we abandon Extreme Project Management and use Agile approach instead? The answer is no. There are situations where the concept is quite suitable, especially in promoting innovation, which is the main theme of this book. Such use cases will be defined and effective tactics for practicing the Extreme approach will be presented in later chapters of this book.

Agile Project Management

Since Agile Project Management evolved from the software industry, let's examine why it provides benefits to software development. A typical software development project usually includes key components such as user interface design, database design, data analysis and report design. In Traditional Project Management, a complete plan is needed before execution starts. Such a plan should include how the user interface looks, how the selection of the database with tables, fields and data relations is defined, how the data is analyzed and how the data reports are presented. The plan would typically be presented to users and management for buy-off and then it would be executed. If the plan is accepted by users, the design and coding are done next. Keep in mind that users are typically not experts, which is why they ask experts to do this project for them. Even though the details are documented in the plan, it is difficult for users to imagine what the system will actually look like when it is done so it is likely that users will not be completely satisfied after seeing the actual system.

Some project managers involve users throughout the duration of a project. This has its drawbacks, however. For instance, after seeing the user interface design, the users might want to add an entry field to collect data that were not initially considered. That would mean the project plan must be updated with the changes in the database design and report design, and the changes must also be communicated to the affected team members, all of which were unplanned administrative efforts for the project manager. Another scenario would be that one of the team members found a better way to store the data and proposed a change. The project manager would face the decision of whether to implement the change. If the change required significant change to the plan and the benefits were limited, the answer would likely be no. Clearly, changes are not tolerated under such project management setting and additional administrative efforts are needed to manage changes.

Using Agile Project Management in this case would mean first breaking down the project into phases such as user interface design, database design, data analysis design and report design. A complete plan is not needed before the project being executed; only the planning of the user interface design phase is needed at the beginning. The completed user interface design would then be presented in a user review session.

If the user wants to add a new data entry field, it would not be a big deal since neither the database design nor the data analysis and report design has started at this point. No additional administrative effort is involved in changing the plan. When the user is completely satisfied with the user interface design, the project moves to the database design planning and then execution.

Similarly, when a team member proposes a better way to store data, the change would be easily implemented as the data analysis and report design has yet been planned. Obviously, Agile approach embraces change by viewing change as a progress to a better solution and continuously adapts to the project situation for better user satisfaction. It also minimizes the administrative effort required to manage the project while simultaneously encouraging team members to learn by doing and discover better approaches to improve the project outcome.

If Extreme Project Management were used to manage such software development projects, the project manager would typically be overwhelmed with numerous changes from users and team members. Because everyone is free to move in a direction that they prefer, that would usually lead to an increase of project scope and size. As a result, integration is extremely challenging. The effort needed to pull everyone together to achieve the desired result would be phenomenal, and only a few project managers possess such strong leadership skills to be able to pull it off. With practically no existing plan, the schedule and budget are ill-defined, which also makes the project very difficult to manage. Therefore, software project managers often choose Agile Project Management.

With Agile Project Management becoming more popular, many variations have surfaced. Wysocki categorizes them into iterative project management and adaptive project management [7]. The best representation of the iterative approach is the Scrum framework. It breaks the project workflow into fixed duration cycles called "sprints." Each sprint starts with a planning session and ends with a review and retrospective session. In between is a work execution with daily Scrum sessions. The daily Scrum is a stand-up meeting for development team members to discuss what they completed yesterday, what they plan to do today, and what potential impediments could impact their progress. Scrum is also a very well-structured approach with concrete rules as to how the sprints are managed and the roles of the participants: Product Owner, Scrum Master and development team. For large projects, the Scaled Agile Framework (SAFe) was developed to integrate multiple Scrum teams with the concept of Agile Release Trains. It combines the Scrum framework with Lean product development and apply cadence and synchronization to schedule management [8].

Adaptive project management is represented by the Adaptive Project Framework (APF) developed by Wysocki himself. Wysocki states that in the iterative Scrum approach applies the learning and discovering methodology within the team itself. The proposed APF is designed to allow customers and users to participate and provide feedback to the project team with a client checkpoint included before exiting each cycle [7].

Undoubtedly, the Agile Project Management approach is currently the emerging approach with fast growing supporters and practitioners. Many companies push the concept in their organization and certainly there are many success stories. However, there is also significant push back from engineers and developers since these approaches are so structured to the point that many people feel very confined and even demotivated. In later chapters of this book, the Agile Project Management approach will be used but once again, the use cases will be defined and effective tactics for practicing the Agile approach will be presented with less bureaucracy and fewer rules.

Summary of the Project Management Approaches

Since we have many different project management approaches available, the question is deciding which one to use. The developers of these approaches and their advocates naturally promote their concepts as the best solution for managing projects. However, these approaches should not be viewed as competitive alternatives; it is not about which one is better than the others. All approaches are effective if they are applied to the right projects under the right circumstances.

In his book, Wysocki uses the model with two scales, "goals" and "requirements & solutions," to classify the projects. He suggests that Traditional Project Management is suitable for projects with clear goals and clear requirements and solutions. Agile Project Management is best for projects with clear goals but unclear requirements and solutions. Extreme Project Management is appropriate for projects with unclear goals, requirements and solutions. I disagree with such project classification. As mentioned in earlier sections of this chapter, projects should have significant importance and impact, as they are the strategic endeavors for the future of an enterprise. As such, why does a company use resources doing projects with unclear goals? Is it just an attempt to hit something big with luck? In my opinion, every project done in an organizational setting should have at least one or sometimes even multiple goals, and these goals must be clear to the project team to be effective to achieve desired results.

The next chapter will propose a different method for categorizing projects so that the suitable project management approaches can be applied to the right type of projects.

Chapter 2

The Project Management Field

Exercise Questions

E-mail your thoughts on the questions to <u>MBPM.Innovation@gmail.com</u>
I will share my thoughts and answers to the questions.

Rules: (For details and reasons, please read Preface Page xvii)
1) One question at a time and state the Question # in the subject line of the email.
2) Provide a scanned copy of the book purchase receipt the first time you use an email to send in a question. This won't be necessary for future questions using the same email.
3) My response will be sent to you between 2-4 weeks after I receive your email.

Q2-1. Should a company balance resources applied to project and operational activities? If so, what would be the criteria for the resource allocation?

Q2-2. With innovation being the target of project management, what should a company do to build such objectives in its strategic planning?

Q2-3. Since statistics show that it takes many projects to make a market success, should companies implement measures to better select projects, so a higher success rate can be achieved?

Q2-4. With the "control" project management system, management keeps the project execution within boundaries and makes those boundaries clear to employees through appreciative and corrective actions. Why are these actions ineffective, especially when giving out rewards as appreciation for employees who achieve good performance results?

Q2-5. What do you think enterprises in traditional industries should do to improve project management effectiveness while Traditional Project Management approach is required?

Q2-6. In your opinion, what are the main issues that companies encounter when they are trying to implement Agile Project Management? Do you think management should continue to push for agile implementation regardless of the push back from their employees?

Management by Project Mapping

The Concept of MBPM

Management by Project Mapping (MBPM) is a strategic concept for sustainable corporate innovation.

Innovation depends on the capabilities of a corporation.

The capabilities of a corporation are built upon its corporate culture.

The culture of a corporation is hindered by its corporate system.

Foundation of Sustainable Corporate Innovation

A company can strategically use project mapping and the right project management tactics to:

- establish an agile, balanced, and open organizational system,

- build a committed and collaborative corporate culture,

- create a capable organization with strong desires for challenges and mastery, and

- continuously achieve breakthrough innovations.

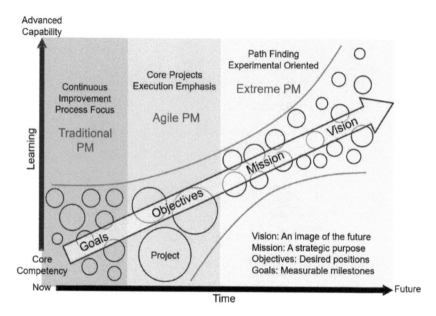

**MBPM as a Strategic Model for Sustainable
Corporate Innovation**

MBPM maps various projects into three areas: Path-finding, Core and Continuous Improvement. In the figure above, circles represent projects and the size of each circle represents the size of that particular project.

Path-finding (PF) projects should align with the company's corporate vision and mission, aiming to obtain the necessary capabilities needed a few years in the future. Core projects should either be the next big product or service offerings to the market, which are needed for the company's continuing survival or market domination. Continuous

Improvement (CI) projects aim to gain the operational efficiency to improve the company's profitability and market position.

Clearly, different approaches should be used to manage these different categories of the projects. The inhibited risk-taking found in most current project management approaches makes them poor choices for PF projects as they restraint innovation. Different matrices are needed to measure the success of the PF projects as many of them would not materialize in new products or services but result in additional knowledge and capability. Therefore, MBPM stresses the use of the right PM approaches for the right projects. There will be chapters in this book devoted to managing projects in each of the PF, Core and CI project categories.

MBPM also emphasizes the all-inclusive approach, where everyone in the organization participates in projects. The objective is to set up a corporate culture of innovation. If many people in the organization are not participating, such culture would not exist.

To better understand the MBPM concept and the reasons behind the concept, please read the Chapter M3 section. Otherwise, go to Chapter 4 for further discussions on building the foundations of sustainable corporate innovation.

Management by Project Mapping

Strategic Elements of MBPM

Most current project management practices emphasize processes and techniques to accomplish the objectives of an endeavor, usually a new product or service. It is rather tactical. Management by Project Mapping (MBPM) utilizes the characteristics of projects and project management at the strategic level to transform an organization and, as a result, increase project success as well. This strategic use of projects does not interfere with the tactical practices of project management; rather, it creates a holistic design for companies to evolve and innovate continuously.

MBPM has four key strategic elements:

1. Strategically utilize projects to build the foundation for sustainable innovation, from system to culture to capability.

2. Map projects into Path-finding (PF), Core and Continuous Improvement (CI) categories with the alignment of corporate development through time.

3. Use the right project management approaches for the right projects.

4. Be inclusive: projects for every employee to create an innovative corporate culture.

The first strategic element (utilizing projects to build the foundation for sustainable innovation) is about setting up the organization for long-term continuous innovation. When a corporation sets an objective to become an innovative company, it must build the capability and the know-how to achieve results that exceed the expectations of its customers and markets. The capability of an organization first reflects on the talent of its employees, and since technology is advancing, it needs to be a learning organization to gain new skills continuously. The company must thus have the adaptability to quickly react to the changes in the business world.

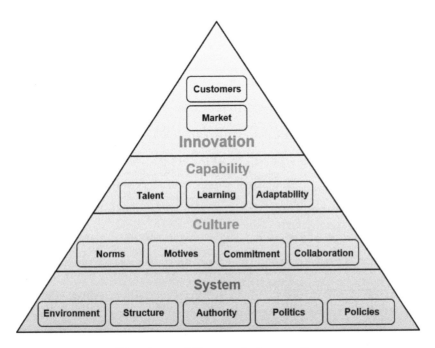

Structure of Corporate Innovation

To become such a capable organization, the organization needs to have a corporate culture advocated by talented, committed employees with great motivation to seek advancement, and who have strong desires for learning and can quickly adapt to changes. Employees must also be willing to collaborate to achieve greatness with everyone making such behaviors the norms of the organization. Such an energetic and vital culture is essential to sustaining and advancing corporate capability.

Corporate culture is built on the corporate system, which consists of the physical work environment, organizational structure, authority, politics as well as policies. An organization's corporate system determines its culture. A company will not have an open and collaborative culture with a closed and bureaucratic corporate system.

In short, the foundation of innovation is corporate capability, the foundation of corporate capability is corporate culture and the foundation of corporate culture is corporate system. The key consideration in setting up the corporate system is therefore determining what culture the corporation wants to have. It is not about business results and performance management. Taking one step at a time, a company should design its corporate system for the desired corporate culture, and similarly, it should build its culture for sustaining and advancing capabilities. With a capable and motivated organization, innovations will arise naturally. Chapter 4 will elaborate on how the foundations are built one level at a time, utilizing project mapping to attain continuous innovation.

The second strategic element of MBPM is putting projects into different categories so they can be managed differently and effectively. As described in chapter 2, the one-size-fits-all approach to project management is not effective. In addition to Traditional Project Management, we now have the Extreme and Agile approaches. Many companies made the logical choice to replace the traditional approaches with the newer ones based on input from developers and advocates of these newer approaches, but they found that success is not always guaranteed. Like many management approaches, they are situational. Therefore, we need to define the situations in which one approach is better than the others.

Many project management practitioners categorize projects by their characteristics, such as clarity of the objectives, requirements and solutions as well as complexity, size, etc. Such categorization is ineffective, as it is done on a tactical level with the examination of the individual project. This book takes a different approach by categorizing projects at the strategic level as part of the company strategic plan to achieve its vision. Projects are classified as Path-finding (PF), Core and Continuous Improvement (CI) projects. Each category of projects has a different strategic importance. Missing or having inadequate quantity of projects in each category could result in significant shortcomings for the company.

The key purpose of PF projects is to explore and obtain advanced skills while generating opportunities to aid the company in reaching its vision. Core projects are the next revenue generating products that establish and enhance the company's competitiveness in the market. Promoting CI projects at all levels of the organization is the key to driving efficiency and establishing an innovative corporate culture. This categorization method actually makes strategic planning easier. Continue to read the L section of this chapter on the details of how the categorization should be done.

The third strategic element of MBPM is applying different project management approaches to manage the PF, Core and CI projects. Typically, Extreme Project Management is suitable for PF projects. Project successes and success rates should not be used to measure the performance of PF project teams. Currently, there is no prevalent extreme model that is widely practiced in industries, even though many models are being developed and proposed. Developing a model for Extreme Project Management is not an objective of this book. In fact, there should not even be a model in managing PF projects as any model could impose a framework limiting creativity. The true extreme approach is not a management model or framework. It is more like a philosophy or style of work. So how should PF projects be managed under this newly defined extreme approach? An entire chapter, Chapter 8, is devoted to exploring the practices that improve the performance of PF project teams to maximize the value of PF projects.

Core projects, on the other hand, are managed very differently. The success of these projects is extremely important, as they must attain fruition with both quality and velocity. Notice that the word "velocity" is used here instead of "speed" as velocity contains the directional vector, whereas speed is just simply a scalar quantity. That means Core project teams must act not only fast but also in the right direction to achieve innovative results. In general, Agile Project Management approach is appropriate for managing Core projects. However, many of Agile Project Management methods are developed with the traditional management mindset and have many rules and a lot of bureaucracy. Chapter 9 is devoted to managing Core projects and will present methods with less administrative efforts to achieve results with quality and velocity.

CI projects are the fundamental ingredients for building an innovative corporate culture, which leads to the fourth strategic element of MBPM: projects for everyone. In most corporations, first line workers make up the largest employee population. Since corporate culture is shaped from the beliefs and behaviors of the majority of its members, there will not be an innovative culture if a company just relies on the R&D departments to innovate and where a large portion of employees do not participate. At best, there will only be a sub-culture within the R&D departments. Participating in CI projects also allows front line employees to learn and improve their skills, which creates the capability by building a learning organization. This ties back to MBPM first strategic element of setting the foundation for continuous corporate innovation.

Therefore, at the strategic level, participation and integration are the key considerations for driving CI projects. On the tactical level, CI projects allow the company to refine and improve its processes to attain noticeable short-term benefits. As such, the "plan the work - work the plan" Traditional Project Management approach is applicable but selectively adopted for managing CI projects. The complex process-oriented Traditional Project Management should not be used since CI projects are usually projects with small scope. Since participation is encouraged, management should allow failure; not all CI projects need to be finished successfully. This is also different from the Traditional Project Management view on pursuing success for every project. Chapter 10 will elaborate on how CI projects can be managed effectively using a modified traditional project management approach.

Chapter 3

Management by Project Mapping

Categorizing and Mapping Projects

Companies often find strategic planning difficult due to fuzziness and uncertainties that they face. The project categorizing and mapping aspect of MBPM is actually a new approach in strategic planning. Before discussing how it should be done, I would like to share the background of how this approach is developed, as it will help to understand the concept.

I have been a career advisor at Intel for many years and have also been advising university students due to my teaching engagements. The MBPM project categorization and mapping concept is derived from the career development model that I use for advising individuals. In the advising sessions, I generally ask, "what is your ideal career image at the peak of your career and when will that be?" Some people know what they want and some do not, which is fine. I explain to the advisees that I am trying to see the vision of their career. Next, I would ask, "what is your next desired job position?" Typically, people can answer this question with certainty. Then I ask: 'What are the things that you can improve to help you be more efficient in your current job?" Building from their answers, I would suggest additional skill improvements. I would like to see them utilize these improvements so that they can be successful at their current job before moving to their next job. Changing jobs because of failing at the current job is usually not the right approach in career development. The three questions that I ask are targeted towards understanding the advisee's long-term vision, mid-term objectives and short-term goals in career development.

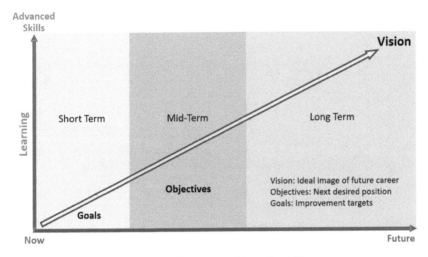

Career Development Planning Chart

Next, I show the advisee the above career development planning chart. It has time on the x-axis and learning on the y-axis. The advisee needs additional skills for reaching the ideal career vision, the next desired job position and the efficiency of the current job, so we list them along the learning axis. Then we work together to build the activities that progressively move up from the current competency to the advanced skills along the learning scale. We map these activities along the time scale in short-term, mid-term and long-term periods. These activities are essentially projects for the individual to achieve their short-term goals, mid-term objectives and long-term vision. Therefore, we need to have activities in all periods. We typically write these activities on the chart as a draft and then transfer them to an Excel format, like the one shown on the next page, for easier modification and better tracking of the progress. The short-term timeline is in months, the mid-term timeline is in quarters and the long-term timeline is in years. Execution starts from the bottom row of the table and moves upward aligning with the career development planning chart.

Vision: Managing technical marketing of key product programs
Target Date: 2021

Knowledge Areas	Jan-17	Feb-17	Mar-17	Apr-17	May-17	Jun-17	Jul-17	Aug-17	Sep-17	Oct-17	Nov-17	Dec-17	Q1 18	Q2 18	Q3 18	Q4 18	2019	2020	2021
HR & Management																			Management Opportunity
Graduate Degree													Explore		Back to school				
Business laws															Book or class				
Relation management													Book or class						
Finance/Eng. Economics									Book or class										
Marketing			Mentor	Book		Book		Book		Book			DOT		Class	Class			
Presentation Skills		Toastmasters																	
Project Management		Book		Class		PM opportunity		Manage projects											
Time management	Book																		
Time	Jan-17	Feb-17	Mar-17	Apr-17	May-17	Jun-17	Jul-17	Aug-17	Sep-17	Oct-17	Nov-17	Dec-17	Q1 18	Q2 18	Q3 18	Q4 18	2019	2020	2021

Goals: 1. Master timing management within 2 months by the end of 2/17
2. Learn project management basics within 6 month and seek PM opportunity in H2 17.
3. Learn presentation skills by joining Toastmasters and in the program through the year.
4. Learn marketing by finding a mentor and complete 1 book bi-monthly as recommended by the mentor
5. Learn finance and engineering economics basics by the end of year.

Objectives: 1. Become a project manager
2. Complete a DOT assignment in marketing
3. Start a graduate degree program

Detailed Career Development Planning Excel Template

As mentioned earlier, some people do not have a clear vision of their career. For those people, I try to understand their passions through a series of "why" questions and help them to build their preliminary visions. We then come up with some exploration activities to confirm their interests. Even with people who know what their visions are, as their careers progress, they may change their minds so we regularly review their career-planning template to make modifications.

Now we can see the strong resemblance between a company's strategic planning and an individual's career development planning. Shown on the next page, the MBPM strategic planning chart has the same time and learning axis as the career-planning chart. Instead of activities, projects are mapped to enable a company to move from current core competency to advanced capabilities over time. The circles represent projects and larger circles represent larger projects.

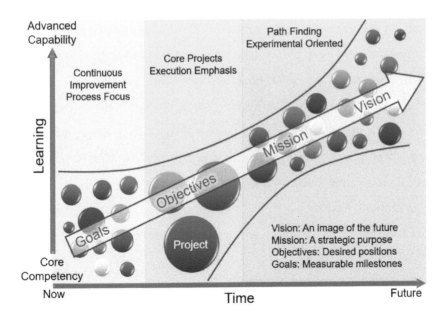

MBPM Strategic Planning Chart

Projects that are aligned with the company's long-term vision and mission are experimentally oriented. These projects are called Path-finding (PF) projects as they are exploratory in nature. The purposes of PF projects are 1) exploring to refine and strengthen the vision, 2) building leading capabilities for the future, and 3) creating options as the next Core projects are coming from these PF projects. A company should therefore plan on many PF projects. The world's leading technology companies, such as Google, Amazon, Facebook, etc., are doing large volumes of PF projects, which allow them to possess leading capabilities, broad market opportunities, and great visions that attract industry attention and talent.

Some people, especially those who work for smaller companies, may question where they would have access to the resources required to do all these PF projects. We will discuss more about resources in future chapters where execution is the focus. In short, people who have such concerns are similar to people who claim that they are so busy with their current job and have no time for future career development. It merely sounds like an excuse. Although luck is sometimes involved, most successful career people start with entry level jobs, just as most large corporations started small. The typical resource driving PF project execution is research labs. Most large corporations have such organizations. Examples are Bell Labs, Xerox PARC, GE Global Research, HP Labs, Intel Labs, Google X Lab, Facebook Creative Labs, etc.

A company should plan for a large quantity of PF projects, each of which is typically relatively small. These projects should be constantly under review so that they are a fit for the company's vision and the desired capabilities the company wishes to attain. Not all projects need to be finished and may be stopped if needed so resources are invested wisely. There is a notion of "fail fast, fail often, fail better and fail forward." Some people believe that this notion contributed to the Silicon Valley success while others consider that it is just hype and that bottom-line failure is a bad thing. Despite the disagreement, this philosophy does generate a more failure-tolerant environment, which is suitable for PF projects. Maybe it could be rephrased as "test fast, test often ..."

Projects designed to achieve the company's mid-term objectives are the Core projects. These projects represent the company's next big market entries, so an emphasis on execution and success is important. The main purpose of Core projects is to develop the company's next cash cow. Although learning does occur during the development of Core projects, it is not the main focus. Core projects are typically large with significant resource investment. As such, a company should have a clear focus and avoid doing too many Core projects. In my opinion, the recent failure of the Chinese company Leshi (乐视) Internet Information and Technology Corp. was mainly due a lack of focus; resources were spread too thin in pursuing too many core endeavors. Core projects should be prioritized over other projects in the company, which should be clear to every employee in the company. Core projects are typically carried out by product development groups and engineering teams.

Projects that are intended for short-term efficiency gains, which are process-focused, are Continuous Improvement (CI) projects. The purposes of CI projects are 1) allowing everyone to participate in projects to create a continuous improvement culture, 2) maximizing the value created by the previously executed Core projects through process improvements and 3) providing opportunities for employee development to increase overall organization health and capability.

A company should plan to do many CI projects that are generally small in size. CI projects do not need to be completed successfully, and failure should be tolerated since the goal is participation. Employees should be encouraged and have the freedom to start and end CI projects. These characteristics are similar to the PF projects but the main difference between CI and PF projects is how they should be initiated. PF projects are derived from the company's vision and generally start from the top by upper management. Conversely, CI projects should be promoted from the bottom and pushed upward. Every single employee in the organization is challenged to examine their daily tasks to propose improvement projects to reduce waste and gain efficiency. The efficiency gains from the CI projects will free up time and resources to fund more CI projects, fueling a cycle of continuous improvement.

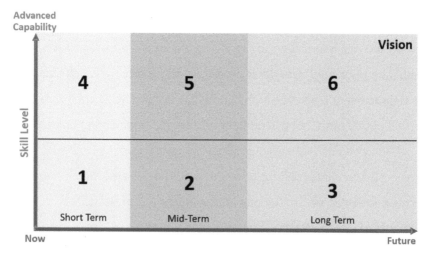

MBPM Project Mapping Chart

MBPM can be used to evaluate the prospects of a company. Pick a company that you know well. List all the projects that are currently being done. Then place these projects in the MBPM Chart above based on their targeted time frame and skill level. The mapping time period is based on the industry and the company's product development cycle. Typically, short-term means one to two quarters, mid-term means six months to a year or two, and long- term means two to five years. When completed, we can analyze the company's performance prospective based on which regions the projects are shown. The preferred pattern would have many projects shown in region 1 and region 6 with a few projects in region 2.

In the long-term period, lacking projects in region 6 means the company would most likely not be a future leader in its industry as it does not explore the leading technologies needed to gain advanced capabilities. Without these PF projects, the company will have limited choices for Core projects in the future, so its market position could be in jeopardy. Projects in region 3 should be canceled as these produce limited benefits for the company. Since the skill requirements associated with these projects are relatively low, they will not provide much competitive advantage since they may have already been done or are easily replicated by other companies. In addition, these projects would not be challenging for the team and result in limited learning. Another alternative is to move them to region 1 so they can be done by regular employees, freeing up valuable research resources to do more projects on developing leading capabilities.

An absence of projects in both mid-term regions 2 and 5 means the company is struggling to find its position in the industry. If the company has many projects in the long-term regions, it is in the discovery phase with a lot of ideas to explore but lacks the one or two big things that are needed to set roots in the market. If this continues, the survival of the company will be in question. On the other hand, too many projects in the mid-term region would mean the company is lacking focus. Strategically, this is not an effective way of using resources. Continuing on this path, the company will eventually overextend itself and finds itself in a resource crisis. A company should prioritize its resources on a few key wins, unless the company is large enough to have abundant resources and multiple divisions in multiple industries. In that case, the division should be the object of analysis, rather than the entire company.

It is preferred to have a few projects with a bigger scope in region 2. It is okay to have projects in region 5, but the company needs to understand that these projects come with higher risks as they require advanced capabilities ahead of their maturity and hence will potentially be very challenging for the team. Often these projects would most likely experience delays, which essentially means they are shifting to region 6 where they probably belonged in the first place. If the company is planning on these projects as the next big wins in the market, then delaying these projects would result in missed market expectations and hurt the company's credibility. Remember, achieving success results is incredibly important for the Core projects. Companies need to take risks but these risks are calculated risks and not foolish risks. In Chapter 6, the topic of risk-taking and risk management will be discussed in greater detail.

In the short-term period, region 1 and 4, if no projects exist, the company does not have a continuous improvement culture and will miss the opportunity to optimize efficiency by not leveraging a large percentage of its employee base. If there are projects in this period but they are done by the development teams or even the research teams, the company typically has a traditional organizational structure with divided groups. In either case, the impact on the company's performance is not immediately apparent. Over the long run, however, the company will have further separation of the R&D and the operation teams, which leads to communication and integration issues and a united corporate culture becomes very difficult. It would also most likely face motivational and developmental issues of employees not in the R&D groups.

Therefore, we would like to see many projects in region 1. It is rare to have projects shown in region 4 but if there are a few, management should pay special attention to them as the people who initiated these projects are willing to take on huge challenges. They might also be delusional but we should let them try. As Steve Job said: "The people who are crazy enough to think they can change the world are the ones who do." Again, it does not matter if these projects reach a successful conclusion or not. The company should promote behaviors that are vital in building an innovation-oriented corporate culture. Management should provide needed resources and connections to enable these employees to go as far as they can. The company should also evaluate how these projects align with the corporate vision and encourage these employees to collaborate with the PF project teams, but not the Core project teams. The Core project teams should be informed but avoid getting excessively involved as they need to focus on the Core projects, which are the company's top priority.

Based on the above discussion, a company should now be able to analyze its current project portfolio and then design its new project portfolio towards the preferred pattern of MBPM. While using MBPM, bear in mind the key characteristics of the three categories of projects, which are summarized in the table on the next page.

MBPM is a general concept designed for common applications. I do not believe that building a comprehensive framework to cover all the specifics will help people practicing it. In fact, it confines people and complicates things. It may feel vague as many characteristics are described on relative scales such as short, long, small, large, a few, many, high, low, etc. I believe that readers are smart enough to apply them based on their situations. Concepts in future chapters are presented in the same manner. I also avoid using the word "framework" to describe concepts proposed by this book as it implies that users should apply the concept within a certain frame.

	CI Projects	Core Projects	PF Projects
Focus	Efficiency	Execution	Exploration
Timeline	Short-term	Mid-term	Long-term
Target	Productivity	Desired Position	Vision & Mission
Size	Small to Medium	Large	Small to Medium
Quantity	Many	A few	Many
Failure Tolerance	High	Low	High
Strategic Importance	Corporate Culture	Business Results	Capability Leadership
PM Approach	Modified Traditional PM	Modified Agile PM	Extreme PM
Executing Resource	Regular Employees	Development Teams	Research Labs

Characteristics of Different Project Categories

Most business models are generalizations of reality and there will be special cases that do not fit well. Without exception, MBPM will not be suitable for every case. For instance, it would not work well in the cases that companies outsource innovation or purchase innovation through acquisitions. Use whatever fits and feel free to modify through practice. Applying the concepts in this book can be considered PF projects and it is important to explore freely – this is part of the learning process. Continue to read the next chapter, which focuses on the most important strategic element of MBPM: Building the foundations, from system to culture to capability, for continuous innovation.

Management by Project Mapping

Exercise Questions

E-mail your thoughts on the questions to <u>MBPM.Innovation@gmail.com</u> I will share my thoughts and answers to the questions.

Rules: (For details and reasons, please read Preface Page xvii)

1) One question at a time and state the Question # in the subject line of the email.

2) Provide a scanned copy of the book purchase receipt the first time you use an email to send in a question. This won't be necessary for future questions using the same email.

3) My response will be sent to you between 2-4 weeks after I receive your email.

Q3-1. There is no single business model that can be applied to all situations, so in your opinion, what type of corporations would see the most benefits from this concept?

Q3-2. If a company decides to implement MBPM, what approach do you think would be effective? Does it need to be starting from the top at upper management?

Q3-3. How does a company integrate the strategic elements from MBPM into its strategic planning process?

Q3-4. If there was a fifth strategic element in MBPM, what do you think it should be?

Q3-5. Do you think there are some projects that are not classifiable? If so, can you provide some examples? What do you think we should do about them?

Q3-6. Should companies encourage employees to participate in PF projects? If so, how do you address intellectual property confidentiality and information security of those projects?

Building the Foundation for Innovation

Corporate System, Culture and Capability

The key to innovation is people – employees with ability and will.

The right corporate culture creates an atmosphere for employees to learn and thrive. Capabilities come from talented people who continue to learn and adapt. Willingness flourishes under a culture created by committed and collaborative employees.

Building the Foundation for Sustainable Innovation

The organization's corporate system determines its culture. Building the foundation starts with the corporate system.

"A bad system will beat a good person every time."

- W. Edwards Deming

So we should construct the system for the people – the desirable employees who the organization wants to attract and keep. Then, we form a culture of innovation on top of that.

"...culture isn't just one aspect of the game – it is the game. In the end, an organization is nothing more than the collective capacity of its people to create value. Vision, strategy, marketing, financial management - any management system, in fact – can set you on the right path and carry you for a while. But no enterprise - whether in business, government, education, health care, or any area of human endeavor – will succeed over the long haul if those elements aren't part of its DNA."

- Lou Gerstner, Former CEO of IBM

A desirable corporate culture attracts talent and stimulates continuous learning, therefore creating capabilities.

"The organization needs to ensure that talent management principles and capabilities are embedded in the culture itself."

- Allan H. Church & Janine Waclawski

How do we build the foundation?

By utilizing project mapping and the right project management!

By design, a company can introduce projects and utilize the right project management tactics to

- provide a suitable work environment
- change its organizational structure
- reduce dependency on key individuals
- create a flexible and agile system
- offer career development opportunities
- develop leadership skills
- promote collaboration and teamwork
- build a motivated and committed workforce
- introduce fun and enjoyment to work
- encourage risk-taking
- simulate the desires for learning and mastery,

and ultimately drive sustainable innovation.

Read the rest of the Chapter 4 to comprehend the ingredients of innovation and how to exploit these ingredients to set the foundation through the concept of MBPM. Once a good foundation is built, we can then shift our attention to the specific techniques that increase our effectiveness in managing these strategically designed projects. The future chapters of this book will focus on the tactical execution management of projects, starting with Chapter 5, which is about how projects are created - with ideation.

Chapter 4

Building the Foundation for Innovation

Key Ingredients for Innovation

If we want innovation, we need capable people who are also motivated to do it. So the key ingredients for innovation are ability and will.

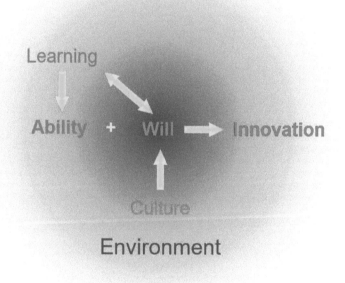

Key Ingredients for Corporate Innovation

In fact, it is not just for innovation. To achieve any success result, we need ability and will. That is the principle that I use for hiring and managing people. I look for people who have ability and will. If I cannot have both, I prefer people who have will over ability. If a person is willing to learn, he or she will eventually gain the ability to a certain degree. Therefore, will is the most important ingredient. We will discuss will first then ability.

A person's will is determined by the individual's personality. It is the most important thing that I am looking for when hiring. Many studies indicated that a person's personality is set at an early age. One of the studies started in the 1960s and tracked about 2,400 individuals for 40 years. It concluded that personality is set for life by first grade [9]. Personality is considered a part of our biology and although it can be altered by life events, such an occurrence is extremely difficult. Therefore, when selecting a candidate for a job position, I want to hire the person who will require the least effort to assimilate. Skills can be trained but personality can't. Caring about a job cannot be taught. In my opinion, my company does not pay me enough to change a person, so I need to pick a person with the personality that fits the job and, more importantly, fits the culture of the organization.

Corporate culture to an organization is like personality to an individual. In fact, the culture of an organization is often shaped by dominating personality traits possessed by the people in the organization. If an organization is filled with employees who love to try new things and have desires for mastery, it will have the culture of a learning organization. Therefore, at the organization level, will is determined largely by its culture. Some may argue that it is determined by the design of the compensation and benefits. Motivation is not mainly determined by rewards. In Daniel Pink's book *Drive*, studies are presented to demonstrate that higher incentives lead to worse performance when the tasks called for rudimentary cognitive skills [10]. Pink concluded that once people are paid enough so that money is not their main concern, the factors that lead to better performance are autonomy, mastery and purpose. All three factors depend on the will of the people, and collectively the corporate culture.

Some management theorists believe that challenging work motivates employees. In theory, this is generally correct, but is difficult to practice at the organization level. How do we determine if the work is considered challenging? It is rather personal. Only the person doing the work knows whether a task is challenging or not. If the task is too hard and seems impossible, the person would not feel motivated to do it. If it is too easy, he or she would not be motivated to do it either. Therefore, at the organizational level, it is difficult for management to set a workload and expect all employees to react with the same level of motivation. The only way to make this work is that managers must have a full understanding of each individual in the organization and then provide tailored work to each individual. Not only is there significant effort involved in doing this, but also this can result in issues related to favoritism and unfairness. Management-driven motivation tactics do not work well, and it is best to use culture to drive the will of the employees.

The other ingredient, ability, is also essential although we prioritize will first. Some abilities can be developed by training given that the people are willing to learn. However, not everyone can gain a particular ability to the same level. For instance, in the case of the public speaking, people with outgoing personalities will have greater success in developing this skill. People who are born shy could spend far more effort in practicing public speaking and still would not get to the same level of success as the former group of people. Hence, the priorities in managing ability are in the order of selection, placement and training. Again, it is the same logic of hiring the candidate who requires the least assimilation effort to perform the job. Clearly, selection is the most important step in setting up the organization. First, we pick people with suitable personalities, which are the basis of will and culture. Next, we pick the people with the right strengths that allow them to develop the desired capabilities easily and quickly.

Once the team is assembled, the main endeavor in ability management becomes training and development. In most companies, management and human resources believe that there is an ideal worker for each job position, so they develop a skill requirement matrix and manage the performance of individuals in those positions against the matrix. The skill competency gaps are then focused areas for development. According the Marcus Buckingham and Donald Clifton in their book *Now, Discover Your Strengths*, it is a flawed assumption to believe that the greatest room for growth for a person is improving the individual's greatest weakness [11]. Buckingham and Clifton believe the opposite – an individual's greatest room for growth is in the areas of his or her greatest strengths. They also believe that a corporation should invest more efforts in the initial selection of their employees.

We can find evidence to support the notion of developing people's strengths versus weaknesses in sports. A TED talk by David Epstein pointed out that athletes get faster, better and stronger with world records constantly being broken. The winner of the 2012 Olympic marathon is nearly an hour and half faster than the winner of the 1904

Olympic marathon. Epstein stated that our human race did not evolve into a new species to warrant such improvements within such a short period. One of the contributing factor is the selection of a specialized body type for a particular sport. He claimed that in the early half of the 20th century, coaches and physical education instructors selected athletes based on an ideal body build for all sports. Now there is an artificial selection for specialized bodies for different sports. In sports where a larger body type has advantages, the athletes got bigger, like American football. On the other hand, in sports like gymnastics where a more compact frame is ideal, the athletes got smaller [12]. Evidently, maximizing the natural strengths of certain body types has helped push forward overall athletic performance.

Just as individuals' physical strengths can be used to achieve greater performance in sports, companies can utilize employees' cognitive strengths to achieve greater performance as well. But what are the strengths for the corporate world? Peter Drucker said, "Most Americans do not know what their strengths are. When you ask them, they look at you with a blank stare, or they respond in terms of subject knowledge, which is the wrong answer." In their book, Buckingham and Clifton offered three tools for people to build their lives around their strengths [11]. They called them revolutionary tools. The first is understanding how to distinguish natural talents from learned skills. The second is a system to identify prevailing talents. The third is a common language in describing the talents. Based on a study done by the Gallup Organization with over two million participants, they developed 34 themes of strengths:

Achiever	Activator	Adaptability
Analytical	Arranger	Belief
Command	Communication	Competition

Connectedness	Context	Deliberative
Developer	Discipline	Empathy
Fairness	Focus	Futuristic
Harmony	Ideation	Inclusiveness
Individualization	Input	Intellection
Learner	Maximizer	Positivity
Relator	Responsibility	Restorative
Self-assurance	Significance	Strategic
Woo (winning others over)		

If you have not read Buckingham and Clifton's book, I highly recommend it. Taking the online StrengthsFinder assessment (https://www.gallupstrengthscenter.com/) may help you to discover your own strengths as well. People with certain personalities and strengths are more suitable for different types of projects that we discussed in chapter 3. Suppose I asked you to identify the appropriate strengths for the team members in Path-finding projects from the above list. What would you pick? Perhaps Activator, Futuristic, Ideation, Intellection and Learner. How about the strengths for the Core project team members?

Clearly, both ability and will, as the key ingredients for innovation, are overwhelmingly serious considerations in the selection phase. What does management then do after the selections are made? In my opinion, besides selecting the right people, another main function of management is to set up the environment for people to grow. Just like in farming, selecting the right seeds is important but it is also important in preparing and maintaining the right environment for the seeds to grow. That environment is the corporate system. If it is not prepared and maintained well, no fruition can be harvested even if there are people with ability and will. That is why Dr. W. Edwards Deming said, "A bad system will beat a good person every time."

In the next section, we will discuss how the foundation for sustainable corporate innovation is built by utilizing the concept of MBPM, which starts with the corporate system then corporate culture and finally corporate capability.

Chapter 4

Building the Foundation for Innovation

Building an Agile, Balanced and Open System

When a person walks into a company, the first impression is based off the workplace's physical environment. The work environment is therefore a significant element of the corporate system. There is no second chance for a first impression and that first impression is based on how people feel. Based on a study done by the Aberdeen Group, about 61% of new employees made the decision whether to stay at the company within the first month of employment and about 26% new employees made the decision within the first week [13]. A month is typically not enough time for employees to understand the corporate culture. That typically takes 6-12 months, so these employees made their decisions based on their first impressions of the corporate system. Often, if you ask the people what factors into the decision to leave, they either cannot describe their reasons clearly or pick on some minor and superficial details. For whatever reason, these people are not fully committed to the company and are just waiting for the next opportunity to show up so that they can leave.

Work Environment

Many companies now understand the importance of the work environment and pay special attentions to designing it. First, art is increasingly prevalent in the workplace. The purpose is to attract artistic individuals, who are both creative and desire mastery – values that companies hope to nurture in their employees as well. More and more companies seek out people who are professionals in a specific field and have artistic backgrounds because they are accustomed to thinking in creative ways and are willing to devote a lot of time mastering skills without focusing solely on monetary rewards.

The "Old" Workplace

A couple decades ago, most workplaces were decorated only using industrial materials with limited dull colors, such as gray, brown and white. It is nearly impossible for an artistic individual to feel inspired when working in a dull place with industrial décor. Subconsciously, artists are not going to be comfortable with spending their entire careers in such an environment. Even those who are not artists are likely to find the lack of color to be drab, and that reflects upon the environment of the company. In such an environment, you may become one of those people who decided to leave the company within the first week of joining. Even if you do decide to stay for some reason, eventually, you will lose your artistic creativeness, and this becomes just a job for you. As W. clement Stone said, "You are a product of your environment."

The Modern Workplace

Most large companies in the Silicon Valley have started decorating the workplace with art. The aesthetically pleasing work environment not only attracts an artistic mindset, but also encourages employees to keep the creative spirit alive.

Many companies also put games in the work environment. Although some think that games are only there for stress relief and team building, they serve other purposes as well. Companies want to also attract game lovers to work in their companies. People who love puzzle games are thinkers and people who love team-oriented games can be competitive, collaborative or both. Like artists and musicians, many gamers spend numerous hours, without sleep at times, playing the games that they love not for money but for the adventure, the advancement of levels and the fun of winning. If you make the work itself interesting and challenging, these people will devote themselves to achieving results that may often exceed your expectations. Again, companies needs to provide the work environment depending on what kind of people they want to have, allowing these people to feel comfortable with a sense of belonging.

Games at the Workplace

The work environment also communicates a powerful message to its employees. A couple of years ago, I visited my friends at Google and one of them told me a story. Google employees often need to go between buildings so to save time, a wire cable was installed such that employees could glide between the two buildings without using the stairs. I could not found any cable between the buildings so I looked at my friend and said that I did not see it. She replied that shortly after the cable was installed, it was taken down since the city officials determined that it might be a safety concern. Why did my friend tell me about this short-lived failed project given the many successful things that Google had accomplished? I felt that she was telling the story with excitement and pride to be a Googler. Many companies encourage their employees to take risks and do so by using "risk-taking" as a slogan or calling it a corporate value printed on posters. Google did it by showing, and it sticks to the employees' minds. It delivered a powerful message to the employees: it is okay to take risk and try. This was a prime example of doing a project not necessary for a successful end result.

Changes in work environment are easily noticeable to employees so companies can utilize projects related to the work environment to promote certain employee behaviors. Many companies have on-going lists of projects to make frequent changes to the work environment, a strong reminder to employees that we are living in a constant changing world, which also sparks excitement from time to time in the workplace.

Many of these projects are based on studies of workplace science. For instance, research done at the University of Texas, University of British Columbia and University of Rochester show that colors in the workplace impact the mood and output of employees [14]. Cool colors such as blue and green inspire creativity and innovation so they are suitable for brainstorming and creative problem-solving. Warm colors like red, orange and yellow are appropriate for action-oriented taskforces and processing meetings since they initiate actions by stimulating the pulse and raising blood pressure. A company needs to have both settings so teams can go to a location that will yield the best results based on the type of work that they do.

Cool Colors Workspaces

Warm Colors Workspaces

Companies should also have both clean and messy work spaces. Studies done at the University of Minnesota and Northwestern University found that a messy space is more conducive to creativity [15] [16]. Allowing mess, but an organized mess, promotes carefree and innovative thinking. On the other hand, a clean and organized work space communicates discipline so it is suitable for execution as the use of the 5S, a visual control methodology for manufacturing with the 5S standing for Sort, Set, Shine, Standardize and Sustain. Separating clean workspaces and messy creative spaces would be optimal for switching between two work modes – messy for ideation and clean for implementation.

Messy and Tidy Workspaces

Office lighting also has an effect on productivity depending on the nature of the work. The Journal of Environmental Psychology examined six studies done on creativity levels related to bright lighting and dim lighting environments and found that low lighting lets people feel less confined and easier to explore and take risks [17]. So dim lighting is suitable for creative thinking; however, bright lighting is necessary when executing those ideas. Poor lighting leads to eyestrain and fatigue, and natural light should be used as much as possible.

Workspace with Low Lighting

There are many projects that can be done to change a company's work environment. Certainly, these projects do not directly connect to the company's business objectives and also take resources, but their impacts on productivity are obvious. More importantly, these environmental improvement projects are done by the building maintenance personnel and their participation in projects to innovate the workplace stimulates the culture to the entire company. A company's environment has significant impact on the culture of the organization, which will be discussed in the section starting on Page 86.

Organizational Structure, Authority, Politics and Polices

Let's look at how projects can be designed to change the organizational structure of a company. Companies hire managers who can grow their businesses. Most managers who are successful in their assigned roles tend to be aggressive and desire domination. The drawback is that they would like to build empires to reduce the possibility of any threat. It is human nature for people to seek job security, power and the feeling of being important. But if an organization does nothing and lets this take its course, the organization will devolve into cliques over time. Many of these groups are efficient in doing what they do so the overall organizational performance may be fine, but the corporation has so much dependency on these key groups that it is practically at their mercy to survive.

A typical scenario happens in many organizations. A business owner hires a general manager to start its business in a new market. So this general manager works very hard and successfully enters the market, establishing a business unit that eventually accounts for a significant portion of the company's revenue. This was exactly what the owner initially wanted, but now the owner is totally dependent on the general manager. Of course, this general manager is smart enough to understand the significant bargaining power he or she possesses and would be foolish not to take an advantage of it. Even if the general manager is nice enough to not make any threats like leaving the company and taking the business elsewhere, it would still be typical to demand more control and independence to strengthen his or her power. Afraid of upsetting the general manager, the business owner has no choice but agree to demands and the situation may deteriorate.

Many companies use the tactic of rotating senior executives to different groups to cope with this problem. It is not a good approach since it interrupts business even if the transitions are well planned. It is like intentionally breaking your child's leg so that he or she cannot run too fast out of your control. This approach also introduces tons of politics as executives want to have control over what their next jobs are and will do whatever they can to not be stuck in undesirable positions. Also, they would not be fully committed to their current job knowing that he or she must give it up in the future.

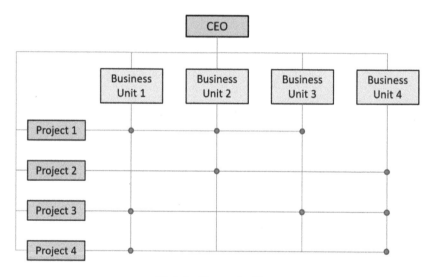

Matrix Organization

This issue is best addressed by utilizing projects. One way of doing this is to map a list of projects across many business units and do them under a matrix organization setting, which is shown in the illustration above. Project managers are given the permission to enter business units and utilize resources so the business unit managers do not have total power to keep their units closed. By working with a given business unit, a project manager is learning how the unit operates while also building rapport with the employees in the unit as well as its clients. This creates potential for this project manager to become the future leader of the business unit. Not only is the power of the managers balanced, but an open organization where information can be easily shared among business units is also created.

The matrix organization also helps balance the stability and mobility of the company. The business units are relatively steady so the company can have a solid foundation to grow. The project side of the organization should be constantly under review and change, which introduces a dynamic element to keep the company vital. Again, for many projects, achieving successful end results are secondary, so the success of every project should not be required since these projects already serve a strategic purpose in the organization setup. Of course, we can aim to achieve both objectives.

Going back to the earlier example where a general manager gained tremendous influence over the organization, the business owner could change the situation by starting a project with a convincing purpose, such as expanding the business into another geographical location. He could then pick one trusted person as the project manager and inform him or her that the main purpose of doing the project is to learn the operations of the business unit and build relations with the key people who work for the general manager. It would be great if the success was duplicated in the new geographical location but regardless, doing such a project would force the general manager to collaborate and share resources, authority and power. After a certain period, the project manager could potentially replace the business unit general manager. The company thus reduces its dependency on certain individuals.

In some cases, when a manager takes over a new group, he or she needs to bring trusted friends or find ways to quickly build trust with a few people. The degree of friendships at work is, however, often difficult to grasp and can lead to over-complicated politics. Some people take advantages of their friendship with managers so managing their performance becomes difficult. Many managers are afraid to make friends at work for this reason. Where there are people, there will be politics. Politics at work has the negative connotation but since it exists, we must deal with it and not avoid it.

Managers can once again utilize projects to address the issue. For instance, the manager can initiate an efficiency improvement project then put his or her friend on the project reporting to a project manager who does not have a close relationship with the individual. With the added management, this friend of the manager now needs to work harder to show performance on the new assignment. Of course, this individual may go to the manager and express his or her discontent with the situation and the manager should be smart enough to emphasis the importance of the project and support the appointed project manager's authority. If that is still not working well and the project manager has difficulty in managing this individual, expanding the project scope and adding another project manager to increase the organization level further will separate this individual from the group manager. Eventually, this isolation will push the individual to either increase performance or leave the position.

Many organization and management theorists advocate a flat organizational structure but in my opinion, the leveling of the organization is not the main concern, but rather, the flexibility and agility of the organization. The above case demonstrates how added levels of the organizational structure can be useful at times. Since projects are temporary, we can utilize this characteristic to create organizations or disband them after the purposes are achieved. In the above case, the efficiency project can be terminated once the performance of those individuals are addressed. The organization could change back as before or take a new form with new projects.

Another disadvantage of a flat organizational structure is the lack of opportunities for advancement. There are few manager positions available, so people feel stuck if those managers do not move up or leave. However, there are unlimited projects that a company can start so there are practically unlimited project manager positions available. Some people may be concerned about the resources needed for doing many projects. We will discuss the cultural aspect of doing many projects in the next section, but the basic idea is that if someone is very passionate about an idea, he or she will pursue it even if resources are limited. If the right culture is present, as we discussed earlier in the M section, the "will" is determined by the culture. With the right culture, employees often need encouragement and psychological support from management more than actual funding. History is filled with successes of under-resourced teams. For example, the Wright brothers beat the more well-funded experts lead by Samuel Langley in inventing the airplane. Allowing employees to start projects and being the managers of those projects gives them the opportunity to dream big and aim high. If it is not required that all projects be successful, what prevents them from trying? If these projects become big hits, resources will come, and they could become the managers of big organizations with huge career advancement. All big businesses start small.

Policies are part of the corporate system and are developed to govern employees. Certain policies in areas like ethics, harassment, safety and security are necessary, but many companies simply have too many policies. A firm should have as few policies as possible. Policies should not be the primary means for managing people; rather, a company should take a cultural approach to managing performance. Projects are once again the significant building blocks of this cultural approach.

Establishing a Willing Corporate Culture

Edgar Schein defines culture as a pattern of basic assumptions shared by a given group. These assumptions reflect from espoused values with observable artifacts [18]. The study of corporate culture was popular starting in the 1980s and again in the millennium, when many new models were developed. Among them are the Harrison model, the Deal and Kennedy model, the Schneider model and the Cameron and Quinn model. These models present different ways to categorize and analyze the culture of a company. Understanding these models can sharpen one's acumen in recognizing the cultural type of a given company, similar to the sense in identifying the personality type of an individual. Knowing the suitable type of culture for certain businesses is important, but it is more beneficial to understand how to use culture to manage performance.

What exactly is the cultural approach to managing performance? It is actually using common norms set by the group to promote desirable behaviors instead of using regulations and controls. I use an exercise in my classes to demonstrate what it looks like. I separated the students into two roughly equal groups. The exercise has two parts and for the first part, I asked the first group of students to raise their hands. Typically, students just raised their hands while sitting and some did not even do that. Then I asked them to raise their hands as high as possible. Most of them did try to raise their hands higher but still sitting. Next, I asked them to raise a little bit higher and most of them tried to raise higher by straighten their arms but most of them still sitting.

For the second part, I introduced the second group to join the exercise and asked this group of students to raise their hands as high as possible. Before doing the exercise, I secretly told about half of students in the group to stand up and stretch their arms to raise their hands as high as possible when I asked. Some of the students who were not told to do so also started to stand up and raise their hands as high as possible. There were a few students reluctant to participate, but students who stood up would naturally look around and eventually those students who were still sitting became the center of attention. Ultimately, they felt uncomfortable and stood up to conform.

The first part of the exercise demonstrates that when management asks their employees to perform, most people will not put in their best effort. Evidently, when I asked people to raise their hands as high as possible, they still would not do it, because many people raised their hands higher when I asked them to raise a bit higher again. This is a typical approach for management to repeatedly push employees to perform, and employees respond by putting in a bit more effort incrementally. The second part shows the cultural approach, where the expected behaviors are set by some of the group members and others feel obligated to follow if they want to remain part of the team. These norms do not need to be set by a majority of the team members.

Sometimes a few respected members of the team are enough to establish the culture of the group. Once the culture is instituted, members who do not conform will feel uncomfortable and usually modify their behaviors to fit in or decide to leave the group.

Now, let's look at an example at work. Management wants to improve the safety awareness at work. Traditional approach would start by communicating the importance of safety at work and asking the team members to attend safety training classes such as hazardous energies, chemical handling, electrical safety, First Aid and CPR, blood borne pathogens, confined spaces, respiration protection, etc. Typically, the progress is slow so management would hold more meetings and send out more emails to push people to take the training. The progress would still be below the expectation of management, so management either provides incentives and rewards for the people who attend the training or makes the training mandatory. Finishing up the classes becomes a formality and a burden.

The cultural approach in this case would be first to find a few enthusiastic and outspoken employees to take the classes and then ask them to talk about the learning from the classes in their daily conversations with others. Of course, managers also need to be part of the program and role model the desired progress. At the same time, management creates a big poster with a matrix of the team members versus the classes and posts it in a high traffic entrance where the team members pass thru many times daily. Classes taken by each team member are updated with star stickers or smiley faces and color code each individual based on overall progress made. Holding stand-up meetings (the meetings do not need to be just for this purpose and can utilize existing process meetings) regularly in front of the poster and allowing a brief time for everyone to look at the poster.

People who have fewer classes completed will feel pressure to increase their participation. Encouragement is also given to ask if anyone needs assistance and emphasize the importance of crossing the finish line together as a team. This approach will yield a better outcome with less management effort. The cultural approach to managing performance is far more effective than the traditional management approach through reward and control. In the future chapters, many new project management practices are introduced based on this cultural approach to achieve continuous innovation.

Don't underestimate the power of culture. Many Silicon Valley companies use culture to attract great workers. Based on a study done by Liz Pellet of Johns Hopkins University, 95% of candidates believe culture is more important than compensation [19]. People want to work with great people and great people build great culture that attracts even more great people, so the virtuous cycle continues. If a company uses salary and benefits to attract employees, when other companies offers more salary and benefits, those employees will leave. Building the right culture leads to a better chance of keeping the desirable employees.

Our generation works hard so that our children will live a better life and do what they enjoy rather than taking a job just to make a living. In some areas of the world, we have achieved this and our children do not need to go into a field just for the money. In China, many companies now have difficulty in managing the youth who were born after the

1990s and even more so those who were born after 1995. They have a different work ethic as they are most likely the only child in the family and well-supported by their parents and grandparents. They are working not because they have to but because they want to, so how they feel about the company is extremely important. Often when they are asked by management for the reasons for doing or not doing something, their answers are generally vague and often include "I don't feel like doing that." One's feelings are a difficult thing to debate. These feelings first come from the work environment, which is most visual part of the corporate system, then from the rest of the components of the corporate system, and lastly from the corporate culture. So the only way to motivate them is to modify the corporate system then build the right culture on top of that. What faces Chinese companies when it comes to managing the single-child youth is an extreme case, but to a lesser degree, it applies to many well-developed areas including Silicon Valley.

A great corporate culture is built on the foundation of the corporate system, which can be changed through mapping projects as discussed in the previous section. I stated that there is a cultural aspect for doing projects to change the system. This cultural aspect ties to the all-inclusive element of the MBPM concept: allowing projects for every employee to create a corporate culture of innovation.

In most companies, projects are done by key functional teams such as research and development so not all employees participate in projects. For instance, in a typical company, the building maintenance crew is only responsible for repairing and maintaining the facilities. In companies like Google and Intel, a large percentage of employees are doing projects. Most of the work environment improvement projects that we discussed in previous section are done by the building maintenance teams. They do research to bring the latest and greatest environmental sciences to the workplace. They do not sit around when the facilities are operating normally; they seek out the needs of their customers, the employees of the company, and strive to exceed their expectations. The previous story about putting a wire cable between the Google buildings for employees to cross over is an example of a diligent attempt to not just doing the job but to be great.

Most of the projects that make Intel a great place to work are managed by administrative assistants. Again, they don't rest after their normal support duties are done. They form teams to bring events and fun to the work place regularly and sometimes surprisingly. For instance, the art decorations displayed at Intel are from local non-profit artist associations and the artwork changes quarterly. Some are from employee photo contests. Intel also has arts and crafts fairs twice a year and occasional music concerts by employee musicians. All of these are low cost measures of introducing art and creativity to the workplace organized by the support personnel staff.

Imagine an employee coming from a traditional company where the support personnel are just doing their assigned jobs to a new company where these employees give 110% effort to provide a better workplace for everyone. How can this employee not work hard? That is the power of culture. If the individual wants to fit into the team, he or she must do what the majority are doing and if not, then the team will not accept this person as a member of the team. Whether or not he or she has management acceptance (by hiring this person) is secondary.

Obviously, the effective use of this cultural approach depends on the openness of efforts visible to everyone. That is why the MBPM concept demands a large quantity of continuous improvement projects so that everyone can participate. These projects allow them to collaborate, from low to high level, from small to big teams, and from technical to non-technical. Through collaboration, their efforts became more revealing and further motivate more employee to join the effort. Culture is a collective mindset of the people and is built upon the willingness of employees to achieve continuous innovation.

Creating Leading and Sustainable Capabilities

After establishing a great culture with motivated people who are willing to put in the effort, the next focus is to create the leading capabilities and sustain the technical leadership. A capable organization consists of three key components: talented employees, learning ability and adaptability. The right corporate system and culture play an important role in attracting talent, but that is only the foundation. Leading researchers want equally competent colleagues and a world class research environment. So to attract great talent, a company needs to build a strong team with selected key target hires. The company must also invest in equipment and lab infrastructure to enable leading development facilities. When resources are limited, cooperating with universities is sometimes an option. More approaches will be discussed in the chapter on managing PF projects.

Learning reduces fear and encourages risk-taking, which is one of the key attributes for innovation. As I mentioned before, I have been doing career advising for many years and I often point out that everyone has a comfort zone and a fear zone. Personal development means reducing both the comfort zone and the fear zone to widen the learning zone. It is intuitive that learning can reduce fear, because the more you know, the more control you will have over uncertainties. How does learning reduce the comfort zone? It is because the more you know, the more you know what you don't know. Once you get out of your comfort zone and look at the situation from a different angle, you will learn that what you believed to be safe before was often just a false sense of security. That is why Intel's former CEO and Chairman Andy Grove said, "Only the paranoid survive," and used this phrase as the title of his book. Learning is therefore for both growth and survival. The same principle applies to companies so we need to build learning organizations.

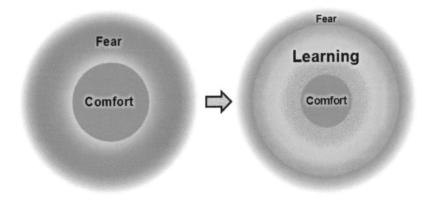

Creating a Learning Zone

To become a learning organization, the organization must first value learning, especially continuous learning and development of its workforce. However, many companies focus on employee training as the main tactic for achieving learning organizations. Some set up corporate universities and bring courses to employees, others offers tuition reimbursement and encourage their employees to take degree or non-degree courses from universities. This approach helps but is not the most effective. The learning ability of an organization is the ability of the organization in gaining the advanced skills in its industry to develop leading products and services. Industry leading capabilities hardly ever come from classrooms. As Benjamin Franklin put it, "Tell me and I forget. Teach me and I remember. Involve me and I learn." How do we involve employees? We do that by strategically mapping projects and encouraging employees to participate in those projects as hands-on development opportunities.

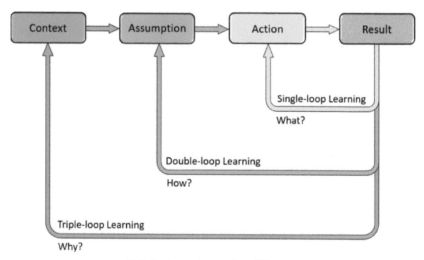

Triple-loop Learning Theory

There are different levels of learning. Chris Argyris and Donald Schon developed a learning theory with the concept of single-loop, double-loop and triple-loop learning [20]. Single-loop learning is about learning what the rules are and the outcome of the learning is knowing how to follow the rules. Let's use teaching children addition: 1+1=2, as an example. Learning is completed once they memorize it and are able to do it again without mistakes. This is the single-loop learning. Double-loop learning is about learning to modify the rules and apply them appropriately. Continuing with the above example, double-loop learning is demonstrated by the children who now can do all simple additions, such as 1+1+1=3, 1+2=3, etc.

Triple-loop learning is about learning the rules themselves and the context of the rules. It is also referred as "learning how to learn." In the same example, we could ask the children to cite a real life incidence of 1+1=2. They probably show adding one pencil to another pencil becomes two pencils. Now, what if we put a feeder fish in a fish tank with an Oscar fish, are we going to have two fishes eventually? Most likely the Oscar fish will eat the feeder fish and only one fish ends up in the tank. In this case, 1+1=1. In another case, what happen if we drop one drop of water on top of another drop of water. Do we have two drop of water? Again, we have 1+1=1. In a different situation, with your mom and your dad, they got together to form a family and maybe ended up with a family of three or more if they have children. After exploring the different circumstances, the children will truly master the concept of addition and understand the conditions regarding when it works or when it doesn't work. Obviously, triple-loop learning often requires life experiences beyond just knowing the concepts, so it is seldom seen in classroom training.

If we want this deep level of learning, we must engage in hands-on real-life projects, which provide meaningful and challenging work for everyone and hence encourage the desire for mastery. It is important for a company to have a learning roadmap, which is done by mapping projects along the learning axis so that there are no gap between current competencies and advanced capabilities needed to achieve long-term vision. Step by step over time, the company will gain in-depth practical skills through doing these projects.

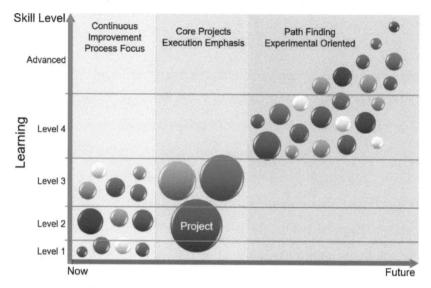

Mapping Projects along the Learning Axis

Many companies have acquired talent by hiring specialists in their field. Many companies also have learning abilities to continue to develop new technologies in their industries. However, having both talent and learning are not enough. For instance, Navy SEAL teams are well-trained so they not only have skills but adaptability is their way of life, which is the means to a successful end [21]. The complexity of a changing business environment has some similarities to the battlefield. New information and threats emerge constantly. A truly capable organization must also have adaptability, which is measured by how quickly the organization can change to react to market conditions. In the example of Kodak, its capabilities in printed photography were unrivaled as it dominated the industry. It also possessed learning abilities as Kodak invented the first digital camera. However, it lacked adaptability and eventually lost its capability to dominate the photography imaging market.

A positive example is Intel. Without doubt, Intel has had a commanding lead in semiconductor technologies since it was founded in the late 1960s. Its capabilities have been well demonstrated with many inventions from products to process technologies. While mainly a memory product company during its first two decades, with the Japanese and other Asian companies entering the market Intel was able to adapt and transform into a microprocessor company. If it had remained as a memory company, it would have failed and been out of business. As the PC market was maturing and declining, Intel once again adapted and transitioned into the data center business. As of 2016, it dominated the data center business with 98% of world's servers running on Intel ships, a market share that surpassed the peak of Intel PC chip market share. Intel is evolving from a PC company to a company that powers data centers and billions of smart and connected devices.

Learning from Navy SEAL teams, there are 4 ways to gain adaptability: 1) have a change mindset, 2) observe, 3) develop courses of action and 4) set small goals [21]. In an organizational setting, all of these can be enhanced by careful design of projects.

1) The frequent introduction of projects, and often ending projects too (fail-fast), changes the work, the organizational structure as well as the tempo of the organization, which creates a dynamic environment. Employees view change as normal occurrences and accept, anticipate and prepare for it. 2) Participating in projects allow employees to do direct observation, a key step in problem-solving. 3) Developing a

course of action means generating plans for possible scenarios. Current project management approaches only plan for success, which is the ideal scenario. MBPM accepts, anticipates and plans for failures instead of avoiding them. When mapping projects, conceptualize possible responses, such as merging, downsizing, upsizing, canceling and selling off the projects, so depending on the actual progress, be ready to act upon the moment. 4) Finally, MBPM calls for many small projects. In fact, most of the PF and CI projects are small projects. The article "The Power of Small Wins," published in the Harvard Business Review by Teresa Amabile and Steven Kramer, discovered the progress principle: "Of all the things that can boost inner work life, the most important is making progress in meaningful work" [22]. Small projects allow us to quickly see progress, which boost our emotions and motivation. Small wins do matter and success builds success. Also, small failures have less impact on us psychologically and the learning from these failures can be tremendous.

In summary, when designing a project portfolio for a company, the main consideration is not about complementary product offering, market entry strategy or risk balance. It is about systematically building the foundation for continuing innovation. The project portfolio should consist of PF projects, Core projects and CI projects, mapped to take the company from current competency to the advanced capability needed for achieve its vision. Utilizing project mapping to build an agile, balanced and open corporate system, then establish a willing and collocated corporate culture, and finally create leading and sustainable corporate capabilities. Thereafter, innovation comes naturally. When you give people a capability to do something great, especially leading capabilities that only a few can master, they will attempt to use it, show it and play it. Innovation is embedded into the organization and no long just a slogan.

Starting with the next chapter, we will transition to the tactical level of project management with tools and techniques that ensure the effective execution of our strategically designed project mapping. The subsequent chapters are organized to cover topics in managing information, managing tasks, managing time, managing PF projects, managing Core projects and managing CI projects.

Building the Foundation for Innovation

Exercise Questions

E-mail your thoughts on the questions to MBPM.Innovation@gmail.com I will share my thoughts and answers to the questions.

Rules: (For details and reasons, please read Preface Page xvii)

1) One question at a time and state the Question # in the subject line of the email.

2) Provide a scanned copy of the book purchase receipt the first time you use an email to send in a question. This won't be necessary for future questions using the same email.

3) My response will be sent to you between 2-4 weeks after I receive your email.

Q4-1. Smart people are sometimes difficult to work with. Do you see a conflict between ability and personality? If so, how would you address this conflict?

Q4-2. Should a company select diverse personality types in a same project team? If so, how to ensure the culture norms are consistent in the group?

Q4-3. Since selecting the right personality is important, how should a company find out the personality types of job seeking candidates and current employees? Do you recommend the use of the assessment tools such as the Myers & Briggs Personality Types and the DISC Personality Test?

Q4-4. If you are in charge of the physical work environment of your company, could you think of some cost effective ways to bring frequent environmental changes to the workplace?

Q4-5. Many people believe that a corporate culture is set by the founder(s) and maintained by its executive staff. In this book, a company's culture is described to be the norms set by a majority of employees. Do you think there are conflicts between these viewpoints? If so, which viewpoint is more valid? If not, how are these two views connected?

Q4-6. A true learning organization develops its capabilities. What are some effective methods in promoting shared learning?

Chapter 5

Project Information Management

Information Creation and Management

Managing projects essentially means managing people, information, tasks and time. The previous chapter was about managing people, which is strategically designed from corporate systems, cultures and capabilities, in order to achieve innovation. Tactical-level project management starts with building an effective project information management system, as it is required throughout the project lifecycle. Effective information management enables optimal performance results, time saving, as well as systematic learning.

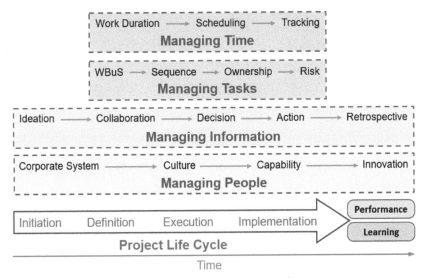

Project Management Overview

Information management has an important role in project management. A project begins with ideation, where information is created. Then throughout the entire project lifecycle, information continues to be generated and requires effective communicating, organizing and processing mechanisms.

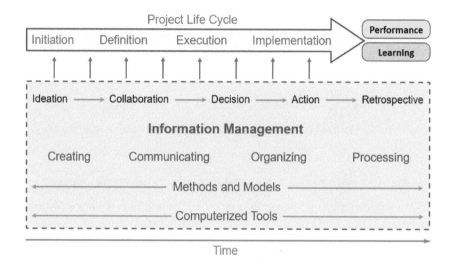

Project Information Management Overview

Brainstorming is probably the most popular technique for ideation. It is simple and easy to use but it usually generates a large amount of low quality ideas. The quality of ideas is important as it decreases the chance of selecting a bad or average idea, and reduces the time and effort required to process them. Two alternatives, Brainwriting and Brainswarming, are introduced and can be applied to yield higher quality ideas with less processing time.

Meetings are necessary for generating, communicating and processing information, but they are one of the most disliked work activities for employees. Effective meeting management is therefore critical for project performance as well as employee morale. The CREATOR system introduced in this chapter will increase the effectiveness of meetings.

In an age of exploding information, we generate a lot of information. Effective sorting and organizing enables us to extract the data needed for decision-making. There are always additional data available for us to consider, so we generally cannot wait to collect all of the information before making a decision. Good decisions are the result of selecting and processing the right amounts of good data. The Six-step Decision-making process along with the RAPID decision-making model help to sort the information and identify the best options quickly. Speed and quality in decision-making can significantly affect the project outcome.

Problems and issues are unavoidable when doing projects. Having a proactive problem-solving mechanism minimizes the impact from surprise elements and provides team members with confidence to face uncertainties. A good problem-solving system includes: 1) having a full understanding of the targeted ideal state, 2) collecting information to fully comprehend the current situation, 3) identifying the gap between the current reality and the goals, 4) generating feasible options and selecting the best course of action, 5) checking the results to ensure the problem is solved, 6) standardizing the solution so it can be applied elsewhere and lastly, 7) developing a plan to prevent this type of problem ever occurring again.

Computerized tools are the high-tech solution that can improve information scalability, accuracy, consistency and security. It has become a habit for most companies to use computerized systems to manage projects regardless of the type and nature of projects. However, using these tools typically required effort to setup, train and maintain. If such administrative work becomes a burden and a constraint to the team, it is better to manage the project with other alternatives, such as using Project War Rooms.

The next section of this chapter (M5) will introduce methods and models for information creation and management, which can be applied throughout the phases of the project life cycle model. The L section of this chapter (L5) will discuss the practical use of these methods and models. Some are developed by me and some are developed by others but with my refinement thought practices. Collectively, these methods and models provide a good tactical foundation for managing projects. Once the information management system is setup, we then transition to managing project tasks (Chapter 6) and project time (Chapter 7).

Chapter 5

Information Creation and Management

From Ideation to Retrospective

Before discussing approaches to project information management, I would like to explain how this book is organized, which is shown below.

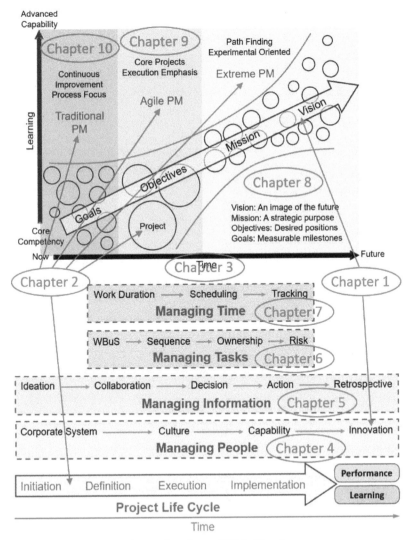

Organization of This Book

In essence, this book can be summarized in two pictures. The top chart is on the concept of management by project mapping (MBPM) shown in Chapter 3, and the bottom graph is about managing projects, which was just presented in an earlier section. Hopefully, it helps you to understand this book conceptually and pictorially. The key theme of each chapter is listed below:

Chapter 1:   Innovation
Chapter 2:   Project Management
Chapter 3:   Strategic Planning
Chapter 4:   People
Chapter 5:   Information
Chapter 6:   Task
Chapter 7:   Time
Chapter 8:   Exploration
Chapter 9:   Execution
Chapter 10:   Efficiency

I did not present this in the introduction or in earlier chapters for two main reasons: 1) I feel that some understanding of the subject knowledge would set the stage so this organization could make better sense, and 2) this chapter is about information management so it would be the best place to demonstrate how information can be organized for better understanding.

This organization reflects my idea of how we should view management in the corporate world. It starts with innovation, which should be the target for most corporations. Innovation is a higher aim than success. The former is achieved by exceeding the expectations of customers and the market, and the latter just by meeting them.

The next management focus is on project management, which is the avenue to achieve innovation. Strategic planning simply means mapping projects to achieve corporate visions. For that, we build the foundation for innovation, from the corporate system to the culture to the capability, which is all about managing people. Then it takes us to the current topic of managing information. Managing people and information are required for the entire project lifecycle, while managing tasks and time cover from project definition to implementation phases.

From a project information perspective, a project starts with ideation. Next, through collaboration and sharing, the project is further defined and takes shape, followed by deciding whether to move forward to execution. Once decided, actions are taken to produce results. After implementation, a retrospective should be conducted to conclude the project by summarizing the performance and key learnings. Throughout the project lifecycle, information is being created, communicated, organized and processed.

Ideation is about generating creative ideas. Brainstorming is the most popular method but it is not effective in many situations. Two alternative methods, Brainwriting [23] and Brainswarming [24], will be explained in the next section. Brainwriting minimizes the unbalanced influence from outspoken people who have strong personalities and is suitable for cross-geography project teams. Brainswarming is appropriate for diverse teams with members from different organizational functions and levels.

Collaboration is about communicating and sharing information, which is a huge topic with numerous methods and models covered by countless books in the areas of teamwork, influence, leadership, negotiation, presentation and many others. Knowledge of these areas will certainly increase effectiveness in collaborating with the teams. However, I do not intend to make this chapter an encyclopedia by reciting all the methods and models in these areas. My desire is to present the practical tools that I find effective while practicing project management. These tools are the CREATOR meeting management system and the SMART Reporting model that I developed, plus the communication instrument "Micromessaging" proposed by Stephen Young [25].

Decisions are the result of processing information. Decision-making is a critical human skill and shapes our lives. Historian Patrick O'Sullivan said, "History is the sum of individual choices and decisions." It has been one of the hottest topics in management studies, which led to the development of many decision-making theories. Many universities offer Ph.D. programs in decision sciences so new theories and models continue to be developed. Once again, it is not my intention to summarize all the theories and models. My objective is to share the practical ones that I personally practiced and find useful, which are the Six-step Decision-making process that I personally developed and the RAPID decision-making model developed by the Bain & Company [26] [27].

Actions are taken after plans and decisions are made, which happen in the execution and implementation stages of the project lifecycle. Project plans and decisions are rarely executed smoothly without problems. There are also many problem-solving models and processes available but most of them solve problems by improving the reality of the current situation. A problem exists when there is a gap between reality and expectations. Problems can sometimes be solved by changing expectations, or more effectively by working from both ends to close the gap. This approach is demonstrated in the Convergent Problem-solving model that I developed.

In managing project executions, utilizing a Project War Room is another popular practice. I believe that using Project War Rooms is a great practice but mainly for managing Core projects. I will briefly introduce this in the next section but will save the details for Chapter 9 where managing Core projects is the main topic.

Retrospective is an exercise done at the end of a project. Under the concept of MBPM, not all projects need to ultimately achieve successful performance but they should yield learning. Retrospective is an important step to capitalize on the learning so that an organization can improve its capabilities. For instance, many projects may not end with a physical product that can be sold, but the processes and technologies developed under these projects may be patentable. Some companies may not want to file patent applications for time and resource considerations, but they can submit their developments for publications. It is a tactic to prevent others from obtaining patents on such developments so that the company is protected from potential IP-related legal issues.

There are no specific tools for this step but only a list of items to be considered. Tools such as CREATOR Meeting and SMART Reporting can be utilized to conduct the retrospective and the RAPID Decision-making model can be used to decide what to do with the developed processes and technologies. In fact, many of the information management tools can be used in multiple steps throughout the project lifecycle from ideation to retrospective.

The discussed methods and models are summarized in the table shown below. The alignment of these methods to the project life cycle phases are in general terms. The purpose of the table is to provide rudimentary considerations for choosing these tools while working in the particular phase. These tools are useful but they are all circumstantial: there is no perfect tool for every situation.

Project Life Cycle — Initiation, Definition, Execution, Implementation → Performance / Learning					
Methods / Models	**Ideation**	**Collaboration**	**Decisions**	**Actions**	**Retrospective**
Brainwriting	✓	✓			
Brainswarming	✓	✓			
CREATOR Meeting	✓	✓	✓	✓	✓
SMART Reporting	✓	✓	✓	✓	✓
Micromessaging		✓			
Six-Step Decision-making Process			✓	✓	
RAPID Decision-making Model		✓	✓	✓	✓
Convergent Problem-solving		✓	✓	✓	
Project War Room		✓		✓	

Key Methods & Models for Managing Project Information

Also, please keep in mind that tools are the means to achieve results. If using tools becomes the purpose itself, you are going down the path of administrative bureaucracy and will stifle creativity and innovation. All tools and models should be applied with flexibility, simplicity and an open-mindedness. I will present how these tools can be used from my perspectives gained by practicing them in my situations. Feel free to modify and refine them through practices in your own situations; using them is a learning process in itself.

The application of these methods and models will be discussed in further details in the next section of this Chapter (L5). For those who want to move on to the next topic of managing tasks, go to the next chapter.

Information Creation and Management

Ideation

We need ideas to start projects. Many people use the technique of brainstorming. While brainstorming is easy to use and most people know how to do it, it has many drawbacks. When generating ideas, the notion of "every idea is good idea" is a false assumption that does not exist. In addition, the quality of ideas generally vary widely, so it takes a lot of time and effort to sort through and extract the good ones. In many cases, people who are outspoken and have strong personalities dominate the discussion and disproportionately influence the idea generation and selection process; people who are shy and reserved may be reluctant to participate.

Brainwriting

The issues with brainstorming can be addressed in many situations with an alternative technique called Brainwriting. A study done in a classroom setting indicated that twice as many students prefer Brainwriting over brainstorming [23]. The technique is known as 6-3-5 Brainwriting or 6-3-5 Method as it was designed to have 6 people, each writes 3 ideas in 5 minutes on a sheet, as shown on the next page, and then transfer the sheet 5 times between the team members. There are also rules such as no talking while writing and writing ideas as short as possible. I found this method useful in many cases but I have to disregard many of the rules. From my previous writings, you probably get the idea that I don't like rules in general as they constrain creativity.

First, it does not have to have 6 people as the team sizes vary at the workplace. Second, it does not have to be 5 minutes for each person to write ideas since some people get ideas quicker, and there is no reason to keep others waiting. Also, there is no reason for transfer the sheet to be passed 5 times if the team is already satisfied with the result. Below is how I would execute Brainwriting.

First, design a sheet as shown below based on the number of people in the team. Although it does not have to do this with 6 people, too many people may not be effective so the prefer limit is less than a dozen people.

	Idea 1	Idea 2	Idea 3
Person A			
Person B			
Person C			

Brainwriting Sheet

Next, pick the first person to write the ideas as shown above. Keep in mind that who you pick to go first matters. If you want creativity and new fresh ideas, do not pick the person with the most expertise in the team to go first. On the other hand, if doing the exercise is to let people feel that they are involved in the decision-making process so that execution would go smoother, having the most senior person in the group go first would set the stage for the desirable outcome.

	Idea 1	Idea 2	Idea 3
Person A	Here is an idea...	...another idea...	...and another idea.
Person B			
Person C			

Brainwriting Idea Writing by the First Person

Then pass the sheet to the second person who has three options available:

1) Write a new idea.

2) Input an idea that builds on an idea from the first person and use an arrow to indicate which idea is used, giving credit to the originator.

3) Combine two or more previous ideas to form an idea and use arrows to indicate which ideas are used.

Continue the exercise with the rest of the team members one by one. Every member has the choice of inputting new ideas, building ideas from others, or combining any previous ideas. Pay attention to the sequence of entries and typically have the most senior person go last.

	Idea 1	Idea 2	Idea 3
Person A	Here is an idea...	...another idea...	...and another idea.
Person B	A new idea...	...an idea that builds from an earlier idea...	...or an idea that combines earlier ideas.
Person C	Continuous to add to an earlier idea...	...combines ideas...	...or a brand new idea.

Brainwriting Idea Entry by Subsequent Members

Once all team members have made their entries, there are a few options to end the exercise:

1) Go to the next round and have everyone review and do the entry again. The original 6-3-5 Brainwriting suggests repeating 5 times but that is overdoing it. Sometime I also allow members to skip or go in different order depending on the situation.

2) Have the most experienced person conclude and pick the final choice.

3) Have the decision maker, often the manager, conclude and decide.

4) Have the team work together to make the decision.

The idea selection process of Brainwriting is much easier compared to that of brainstorming. By looking at the brief demonstration sheet on the previous page, could you identify the best possible choice for an idea? If you pick Idea #2 from Person C, you are correct. The most convergent idea following the arrows would be the obvious choice since many members contributed to it. At a minimum, this choice has less opposition as it is built by multiple members, which leads to easier acceptance and implementation.

We have discussed the preference of the cultural approach to management over traditional management based on structure, authorities, politics and rules. The outcome of a brainstorming session is heavily influenced by individuals with power based on where they fit in the organizational structure and their authority plus the organization's politics. On the other hand, the modified Brainwriting technique presented here has the characteristics that align with the cultural approach to management, which are open, fair and inclusive. The technique allows everyone to participate almost equally since it eliminates the unbalanced domination by individuals with strong personality and authority. It is possible to do this anonymously by posting the sheet in a room allowing people to make entries while others are not present. This also minimize the effects from politics played by certain individuals.

Brainwriting also has another advantage: it can be done without having all team members physically present in a face-to-face meeting and can even be done without holding a meeting. A Brainwriting sheet can be put on the company's network drive, Microsoft SharePoint or, if information security is not a concern, saved as a Google Doc. As globalization continues and many projects are done by global teams, Brainwriting has the advantage in the cases where project team members are not residing at the same geographical location, For instance, the working hours of the United States, Israel and Malaysia are hardly overlaps. Brainwriting allows team members to work on their own time without the need of a meeting.

Brainswarming

Sometimes we are facing issues that requires employees from different functions and levels to come together to work out a solution. For example, a company is experiencing sales decline and management wants to initiate some projects to improve sales. Since this is a high-level issue, holding a brainstorming session would most likely involve people from many different areas. The ideas generated would have a wide variation of focuses and details.

Marketing personnel might have ideas in increasing advertisements and promotions. Engineering might have an idea in customizing products for different users. R&D might have an idea in speeding up the development of a technology and incorporating it into products. These ideas are far apart and would be further complicated when both management strategists and frontline practitioners are involved. The sales manager might have an idea in penetrating into a new market segment while a sales person might have an idea in pursuing a large contract with an existing client. These all seemed to be good ideas, but we need to prioritize and select a few choices in order to create a focused strategy so that the company's limited resources is used effectively. However, since these ideas are so different and not even on the same level of playing field, doing the second part of brainstorming to narrow the choices down to a few solutions would be difficult.

Strategists are typically top-down thinkers who start to think from the goals and then further divide them into smaller goals. Next, the possible actions to achieve the sub-goals are generated and finally the solutions are derived from these possible actions. Frontline practitioners are generally bottom-up thinkers who start to think from the resources available to them and how these resources can be used. Based on the usage modes, solutions are developed.

Top-down Thinkers

Bottom-up Thinkers

Two Different Styles of Thinking

When the two different styles of thinkers get together in a brainstorming session, it feels like they are speaking in different languages with both very general and very detailed-oriented ideas being generated. Brainwriting works a little better as ideas can be built upon and combined with other ideas but convergence is still difficult depending on how far apart these ideas are: there is no clear direction to connect them.

This can be solved with an alternative called Brainswarming, developed by Tony McCaffrey, is designed to accommodate both top-down thinkers and bottom-up thinkers in developing viable solutions [24]. The general concept of Brainswarming is illustrated on next page.

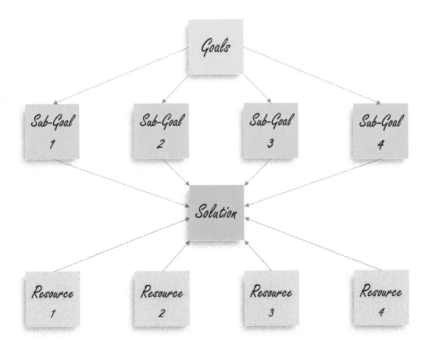

Brainswarming Solution Convergence Flow

I modified and expanded the Brainswarming method by adding possible actions below sub-goals and usage modes above resources. The technique is best done with the team in the same room using Post-it notes in the following steps. First, goals need to be clearly stated. Yes, it can have multiple goals. Next, team members are free to post notes on sub-goals and resources based on their thinking styles. Possible actions for sub-goals and usage modes for resources can also be proposed right after a sub-goal or a resource is posted. Afterwards, team members identify solutions based on where possible actions can be linked to usage modes.

This process can go on for several iterations as team members can add new sub-goals, possible actions, resources and usage modes to refine and develop additional solutions. Strategists and managers tend to focus on the upper part of the Brainswarming layout above solutions and practitioners typically focus on the lower part of the layout, but this is not a requirement. In many cases, a single individual can think both from top-down and bottom up, which is fine. Finally, the team, working together, decides on which solution or solutions to use and prioritizes them based on available resources. The modified Brainswarming method is shown below.

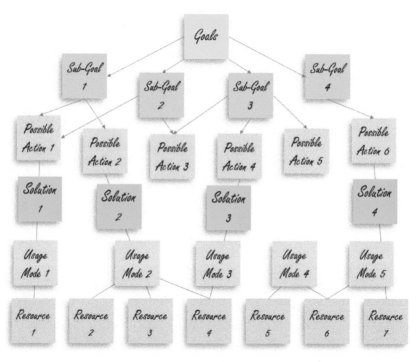

Modified Brainswarming Solution Flow Demo

Clearly, both Brainwriting and Brainswarming are more organized than brainstorming. However, that does not mean they can replace brainstorming in every case. In situations where we want to get many ideas quickly, brainstorming is still effective because the method is well-known and very little explanation and preparation is required. Team members know what is expected from them and can do it fast. Some teams may have established a cohesive culture that addresses many of the brainstorming common pitfalls. In fact, some teams do not use any tools and can generate great ideas by just getting together and talking, which leads us to the next topic – collaboration and communication.

Collaboration and Communication

As mentioned earlier, there are many books in collaboration and communication, offering numerous methods, models and practices. Ideation tools that we just discussed such as Brainwriting and Brainswarming may also be regarded as collaboration and communication tools. Indeed, there are many tools that can improve collaboration and communication at the workplace. I choose to discuss the practices for frequent activities such as meetings, reports and daily conversations, not only from the angle of seeking the best improvements but also from the position of reducing the most frustrations of the team members.

CREATOR Meeting Management System

Attending meetings is not a favorite activity for most employees. In fact, many people hate meetings and feel that they are a waste of time. According to research done by Rick Gilbert, an executive coach for many Fortune 500 companies such as Apple, Cisco, eBay and Oracle, executives consider that more than two-third of meetings are complete failures [28]. Igloo Software surveyed 1000 people and found that 76% of meetings are not necessary [29]. Another survey done by ResourcefulManager on 948 executives and managers indicated that almost half of meetings accomplish nothing [30]. An effective meeting system is important for achieving results and boosting employee morale towards attending meetings.

I developed the CREATOR Meeting management system that helps to increase meeting effectiveness. CREATOR stands for Category, Roles, Expectations, Agenda, Timekeeping, Outcome and Record. CREA are required in the meeting invitation and TOR are how the meeting should be run.

First, the category of the meeting should be clear to attendees. Each company should have a list of common meeting categories. For instance, in his book *High Output Management,* Intel former CEO Andy Grove categorized meetings into process-oriented meetings and mission-oriented meetings [31]. Process-oriented meetings are regular meetings including one-on-one, staff and operational review meetings. Mission oriented meetings are ad-hoc meetings for problem-solving and decision-making. In general, meetings can be categorized as status update, information sharing, decision-making, problem-solving, innovation and team-building meetings. It does not matter on how meetings are categorized: the important thing is to have a common set of meetings that are understood by all employees and then have a standard template for each category of meetings.

Clear roles need to be defined so participants can be prepared before a meeting so as to not waste time during the meeting figuring out who is doing what. Having clearly documented expectations for the participants is important to ensure the meeting is focused and runs like clockwork.

An agenda is super critical for meetings as a large percentage of meetings ineffectively manage time. The survey done by ResourcefulManager shows about 40% of meetings do not end on time. Agenda items must have time allocations as well as owners, so that meeting participants know what is expected from them.

Meeting invites with CREA should be sent to meeting participants ahead of time along with a time and place in email scheduling tool. When is the best time to send meeting invites depends on the category of the meeting and the time required for participants to get prepared. An example invite for staff meetings is shown below.

Category: Staff Meeting
Roles: Facilitator – Lily, backup Nora
            Timekeeper – Ava, backup Jack
            Note-taker – Leo, backup May
            Required Participants – Dan, Ella, Jack, Nora, Mike, May
Expectations:
            - Be on time and prepared
            - Participate: avoid using phones and laptops unless required
            - Share ideas openly
            - Respect others
Agenda:
            8:00 - Safety and urgent issues, Dan/All
            8:10 - Project A update, Ella/Mike
            8:25 - Project B proposal, Jack
            8:40 - Project C requirements, Lily/Ava
            8:50 - Opens, All

Staff Meeting Invite Template

The TOR part of the CREATOR Meeting applies when meetings are actually in session. T stands for timekeeping, which is extremely important for keeping a meeting on track and productive, otherwise the agenda is just a wish list. This key responsibility belongs to the timekeeper who should highlight what time the presenters have before each time slot and remind them when their time are near the end. Timekeepers may use typical sentences such as: "Ella and Mike, you are up and have 15 minutes." "Five minutes remaining." "You only have a couple minutes, please wrap it up."

O stands for outcome, which is another dimension that is critical for running an effective meeting. If we do not get results from the agenda items, discussing them is a waste of time. Keeping focus on the outcome of the agenda items is the responsibility of the facilitator who should ensure a focused discussion with high quality conversations. When there are conflicts and disagreements, the facilitator needs to redirect the discussion back to importance topic so the meeting does not go down a rabbit hole. When people are passive and unwilling to contribute, the facilitator needs to encourage members to share ideas and promotes group discussion. Essentially, the facilitator's role is to ask for facts, opinions and suggestions to ensure the discussion yields the best possible outcome. Facilitators may ask these typical questions:

- "What are we intending to achieve?"
- "What is the meaning of this?"
- "What data supports your claim?"
- "How can we do this effectively?"

R stands for record, which should capture the key points of the discussions and in many cases, the decisions and actions that need to be executed after the meeting. Many people refer to this record as meeting minutes. The note-taker has this responsibility. Proper record of the meeting drives attendance and actions as well as serves as the proof of agreement. Meeting records should be short and simple but also need to capture all the key information and actions needed to be performed. Using the earlier staff meeting example, the meeting record may look like what is shown below.

Staff Meeting Dec. 5, 2017
Attendance: (P - Present, E - Excused, A - Absent)

	WW43	WW44	WW45	WW46	WW47	WW48	WW49	WW50	WW51
Dan	P	P	P	P	P	E	P		
Lily	P	P	P	P	P	P	P		
Ava	P	P	E	P	P	P	P		
Leo	P	E	P	P	P	P	P		
Ella	P	P	P	P	P	P	P		
Jack	P	P	A	P	P	A	P		
Nora	P	P	P	P	P	E	P		
Mike	E	P	P	A	P	P	P		
May	P	P	P	P	P	E	P		

Discussion Summary: (i. - FYI, a. Action required)
- Safety and urgent issues, John/All
  i. Site building shutdown for maintenance on WW50 weekend.
  a. Potential chemical leak identified in equipment C. Need to check all existing tools by WW50 – Leo.

- Project A update, Ella/Mike
  i. Project A progress is on track for completion Q1 next year.
  a. Report out budget spending status in WW51 meeting – Mike.

- Project B proposal, Jack
  i. Project B will improve current market position and attract new customers.
  a. Based on the current proposal, Project B requires about 50% more resources than project A. Inform finance department by the end of this week – Dan.

- Project C requirements, Lily/Ava
  i. 80% requirements are collected and 20% pending. Expected to complete WW51.
  a. Need to double check with customer A to fully understand requirement B. WW51 - Lily

- Opens, All
  i. A cost reduction proposal is being worked by team A. Will bring it to this meeting when ready. Completion date yet to be determined.
  a. Customer B request information on Product C. Need to check with legal department for IP release. WW52 – Ella.

Action Items:

Task	Owner	Done By	Description	Status
49.1	Leo	WW50	Check all tools for chemical leak	
49.2	Mike	WW51	Report out Project A budget spending status	
49.3	Dan	WW49	Inform finance on Project B resource requirement	
49.4	Lily	WW51	Check with customer A on project C requirement B	
49.5	Ella	WW52	Check with legal for IP release on Product C	

Staff Meeting Record Example

In addition, meeting records should be produced in a timely manner. In fact, it would be best to have the note-taker doing it while the meeting is in progress so that it is done when the meeting ends. Writing the record online or using tools such as Microsoft SharePoint and Google Docs allow all the members to see the document as it is being updated so that any errors can be spotted and corrected real time.

To further improve the effectiveness of certain meetings, the following ground rules may be useful:

- Attendance is not required if any part of CREA is missing in the meeting invite.
- Attendees may attend only the agenda slots that are assigned or of interested.
- Cancel and reschedule the meeting when key roles are absent or more than half of participants decline.

Teleconferences are conducted more frequently in recent years due to the increase of global teams and reduction of travel budgets. The CREATOR Meeting system can easily be applied to manage teleconferences by just adding expectations such as announcing yourself before speaking, putting on mute when not speaking, avoiding background noises and no multitasking.

Mission oriented meetings for decision-making and problem-solving will have different templates for meeting invites and records. I will discuss those meetings while presenting decision-making and problem-solving models, as they are the result of merging those models with this meeting management system.

SMART Reporting

Besides meetings, report writing is another activity that employees are not thrilled to do but critical for communicating and learning. Many have probably heard of SMART goal setting, which was coined by George Doran in 1981 and later modified by many others using different words to make up the acronym SMART [32]. The original five words were Specific, Measurable, Assignable, Realistic and Time-related. I adapted the term SMART and developed the SMART reporting with these different words: Straightforward, Memorable, Action-oriented, Relevant and Timely.

Straightforward means communicating points in a succinct and direct fashion. On written reports, key messages should be stated in the first few sentences. Having a summary section in the beginning is also helpful. You can even label it as "Key Messages." For verbal reports, key messages should be in the first slide of the PowerPoint right after the title page or even on the title page. Tell the audience what you are going to tell them, what the important points of your presentation are and then support them with data.

One of my habits is that when I am asked to present a report in a certain length of time, I will develop a version that takes half of the time and another version that takes three quarters of the time. It forces me to be concise and prioritize materials. It is typical that most presentations go over their scheduled time, and in cases when you present the half-time version of the report, you will have the appreciation of meeting participants, especially the chair of the meeting. In a way, your presentation stands out. Under normal circumstances, present the three-quarter version so that time is well managed. There may be some uncontrolled events that take extra time, such as people asking too many questions. If you are on time and no one asks any question, you can summarize the key points at the end so the key messages really stand out for the audience. By the way, the two versions can be built in a single PowerPoint file. All you need to do is to use special marks and graphics on the items that you would skip when doing the short version.

Memorable means making your report standout and impressing your target audience so that they remember what you present. Develop your report from the viewpoint of the audience and make it enjoyable for them by creating an emotional connection with them. Make the presentation more interesting by using visuals but do so wisely. Pay attention to details and do not use hard to read text, bullet points or charts.

Action-oriented means clearly stating what you want your target audience to do and/or what you are going to do. Maximize the value of your report and that value is from actions but not just words on paper. Being a manager for many years, one of my top dislikes is that people report issues without actions, so I have to ask, "What would you like me to do or what are you going to do about it?" Even if the report is all about good news, there can still be some possible actions such as recognizing the team and sharing knowledge across the company.

Relevant means presenting the right information to the right audience. Get the context before you present so you know what your audience cares about. Presenting too much information will overwhelm the audience and cloud the key messages. Too little information will not be convincing enough to establish credibility. Usability is the key measurement for the relevance of a report. The information presented must be useful for the audience, so think about how the content of the report can be utilized when developing it.

Timely means delivering or presenting the report at the right time. Delayed reports delay actions. In fact, management and clients prefer to have access to the project progress in real time, so for written project status reports, it would be best to create real-time reports that users can access at the time of need, or automate these reports so that users can subscribe and have them sent via email regularly. Project managers should consider doing an oral presentation after the occurrence of a major project event, whether it is a planned milestone or an unplanned project problem. Keeping the visibility of the project throughout the project lifecycle is important for the open culture and team learning that we want to achieve and maintain.

Micromessages

Another frustration for people at work is the feeling of being misunderstood or not being understood at all. While we are feeling like victims in the fast changing work environment and complaining that others do not pay enough attention to us, we are doing the same to each other. This is because we communicate far more information with hand gestures, facial expressions and tone of voice than through words. We unintentionally send out a great amount of micromessages through these non-verbal cues. In his book *Micromessaging: Why Great Leadership is Beyond Words*, Stephen Young points out that,

"Developing the skill and language to identify and address negative micromessages puts a new power in your grasp... The new skill will enable you to rally everyone around you with micromessages that inspire, motivate, and get beyond conventional rhetoric [25]."

Negative micromessages are called microinequities. We need to minimize microinequities by being conscious of our facial expressions while interfacing with others. Most microinequities are communicated through facial expressions, which we cannot see ourselves. Many of us can sense signs of impatience, displeasure and absence, which cause us to feel disrespected. We then shut off or cease participating sincerely in return. We also need to monitor our personal greetings and interactions with others. Many people unconsciously show closeness to people who they like and are formal with the others. These signs can be easily sensed, and favoritism is perceived. Responding to differences of opinion is another situation where people easily send out unintentional micromessages. Slight overreaction and subtle signs of disparity are enough to make others feel uncomfortable. All of these microinequities lead to distrust in the work environment that make collaboration more difficult.

Positive micromessages are called microadvantages. To maximize microadvantages, we should first be truly immersed in the conversation through active listening and asking questions. Building rapport with others by sending micromessages that you are interested in them as a person not just the work performed by them. Proactively solicit opinions and at the same time send sincere micromessages that make others feel valued. In team discussions, acknowledge and address others using individuals' names instead of "he" or "her idea." When addressing a group, pay equal attention to every single person with routine eye contact from time to time. When being interrupted, calmly acknowledge the person speaking and politely shift the focus back to the original idea. To learn more about utilizing microadvantages, please read Young's book.

Decision-making and Problem-solving

Ideation is the beginning and collaboration allows the sharing of these ideas to create quality and feasible resolutions. Decision-making then determines which option is the best to pursue. If we could do everything we wanted to do and had everything we wanted to have, decision-making would be simple. Most of us, however, are living in a world with limited resources, so we have to make choices to do certain things while simultaneously giving something up. Author Tony Robbins once wrote, "It is in your moments of decision that your destiny is shaped."

Organizations are constantly facing decisions to undertake certain projects while forgoing other projects. Within projects, there are decisions to be made in choosing resources, processes and schedules. How can we make better decisions? How can we make decisions quickly when many people are involved? For the first question, I will present the Six-step Decision-making process, which provides a general step-by-step guide for individuals and organizations to make better decisions. For the second question, I will outline the RAPID decision-making model, which designates roles in a team environment for fast decision-making.

Six-step Decision-making Process

Shown below, the Six-step Decision-making process involves timing, framing, aiming, forming, affirming and performing.

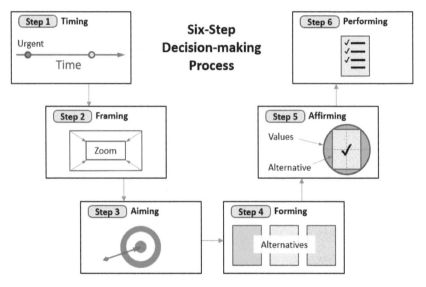

Six-Step Decision-making Process

The timing step is to determine the urgency of a decision. Some decisions only have a small window of opportunity and the consequences will be significant if we miss this window. People who are considered indecisive are in essence missing the best timing to make a decision. There will always be additional useful information that we can collect for a given decision. When to stop collecting data therefore depends on the urgency of the decision. Determining the timing for the decision is the first step, as it defines the subsequent actions in making the decision. A decision that needs to be made today will require a different course of action from a decision that can be made next month. Making a fire-fighting decision reacting to an urgent request, for instance, is much different than making a planning decision for a project starting next quarter.

Framing a decision is the second step, which involves defining the scope of a decision. Framing a decision to broadly may overwhelm us such that we are unable to pinpoint the right issue. Framing a decision too narrowly will lead to missing threats and opportunities. The best illustration for framing a decision is shown in the figure below. Obviously, a bad frame limits our view and prevents us to make the best choice. We need to zoom in and out while analyzing the situation to find the right assumptions for the decision.

| Framing too board | Framing just right | Framing too narrow |
| Overwhelmed | | Missing threats |

Decision Framing Illustration

The third step is aiming, which means setting the general direction for a decision. When we are in a situation where we are not satisfied, there are only three general directions available: change it, accept it or leave it. We should consider making the choice in that order as well. The first choice is to determine what we can do to change the situation. If we can, then the decision is to seek possible actions that alter the circumstances in our favor. If we determine that we cannot change the situation, which is true in many cases, then our next choice is to accept it and then find ways to minimize the impact. We cannot stop the rain but we can use an umbrella. At work, we are often asked to disagree and commit. It means that we need to accept an undesirable fact and work with it. Lastly, if we find that we cannot accept a situation, which we cannot change, then the last option is to leave. In general, aiming is suitable for personal decisions as well, such as working on an undesirable job or mingling in an undesirable relationship. The first attempt is to change the situation. If you cannot, then ask yourself if you can accept it. If you cannot do that either, then the only choice left is to leave.

Once the general direction is determined, we can move to the fourth step, which is forming alternatives. Alternatives are the options that can be acted upon and within the decision makers' power and control. Brainstorming, Brainwriting and Branswarming are tools that can be used to develop alternatives.

	Criteria 1	Criteria 2	Criteria 3	Criteria 4	Criteria 5	Total Score
Weight Factor	1X	3X	1X	3X	5X	
Alternative A						
Alternative B						
Alternative C						

Decision Matrix Template

The fifth step is selecting an alternative that aligns with the decision makers' personal values. A decision matrix with alternatives against selection criteria could be used. An example of a decision matrix template is shown above. The criteria should be aligned to the decision makers' desires and values. Since each criteria may carry a different value to the decision maker, a weight factor can be used to calculate the score for each alternative. The best option to act upon is the one with the highest total scores.

The final step is acting on the chosen alternative and following through the decision. A decision that is not being acted on is merely a wish. We can make everything look pretty in PowerPoint slides showing step one to five but without actions, no value is created.

RAPID Decision-making Model

When decisions are made in a team environment, the RAPID decision-making model developed by the Bain & Company is a good tool to use. Each letter in RAPID represents a role in the decision-making process [27]:

- R is the Recommend role, which does most of the work in the decision-making process. There should be only one R person who pulls all the information together and makes the recommendation to the decision maker.
- A is the Agree role, which validates the recommendation on its legitimacy. The A role can involve multiple people but is generally limited to legal, HR and other compliance and regulatory roles.
- P is the Perform role, which executes the decision. The P role can be multiple people as well but not all decisions need to involve the P role. They can be a good source for information.
- I is the Input role, which provides the expertise and knowledge related to the decision so R can formulate the recommendation. The I role can also be multiple people and they are typically the experts in the field.
- D is the Decide role, which makes the decision. There is only one D, who will be held accountable for the consequence of the decision.

The RAPID decision-making model combined with the CREATOR Meeting model will streamline the decision-making process at work. The category of the meetings will be mission-oriented. The roles in meetings can be simply designated by a single letter from RAPID. The expectation for each meeting can be short and simple, such as determining the D, finding I, Checking with A and so on. Just from these short statements, participants already know who should be at the meeting and what they are expected to do. The R is clearly the person who arranges and runs the meetings. A decision-making meeting should not be held if the D is absent. The team can fully understand why they are there at the meeting. Meeting minutes are also made simple and direct by using these letter designations. The RAPID model enables faster decision-making by clearly defining the roles and responsibilities of individuals.

Convergent Problem-solving Model

After a decision is made, projects move into the execution phase and problems are often encountered. Having problems is not entirely negative. If we are doing challenging work, problems are expected. Therefore, we should have a predefined process for addressing problems, so when a problem surfaces, the team would not be surprised and can act calmly to solve it. A problem exists when the reality is different from the expectation. The bigger the gap means the bigger the problem. To solve a problem means closing the gap between reality and expectation. Most people solve problems by improving the reality. It is more effective, however, to solve problems by working from both ends: improving the reality and changing the expectation. From my experience, the Convergent Problem-solving model is developed.

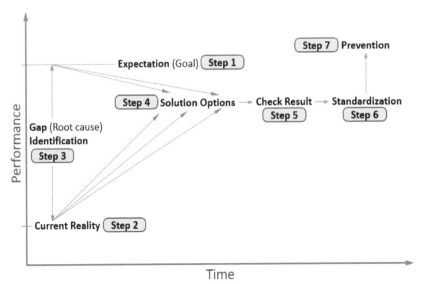

Convergent Problem-solving Model

Step 1 of the model is to fully understand the performance expectations. This is the successful ideal state that will satisfy all stakeholders. Therefore, this is the performance goal that we are shooting for originally.

Since we have a problem, we need to understand the current reality, which is the situation that we are in now. A problem exists when the situation is not ideal for some or all people, so Step 2 is to understand the current situation. If we do not know where we are, it is hard to find the path to where we want to be.

Step 3 is to identify the gap and root causes that lead to the gap between the current state and the expected ideal state. Tools such as the fishbone diagram and failure mode and effects analysis (FMEA) are useful for this step.

Step 4 is to formulate options for closing the gap, which should be looking from both the goal and the current reality perspectives. The similar concept used in the Brainswarming tool is appropriate for this purpose. A solution generation flow chart is shown below.

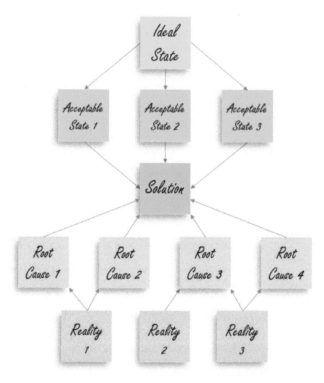

Convergent Problem-solving Solution Generation Flow

From the top-down direction, we need to work with clients or management to define acceptable states, which may be less desirable than the ideal state. It is a process of re-examining the feasibility and achievability of the original ideal state and prioritizing resources for the best-case scenario under the current constraints. From the bottom-up direction, a typical problem-solving approach applies: identifying the root causes of the issues and addressing those causes. When developing solutions, we also need think about the time scale, which means developing short-term, medium-term and long-term solutions to the problem.

Step 5 is to put the selected solutions into action and check to see if these solutions are truly getting us to an acceptable state. Many problem-solving models end with this step. However, if we want to be proactive in addressing similar problems in the future, we need to take this further.

Step 6, which is to standardize the solutions so that they can be readily applied should the problem occur again. This step is especially beneficial for large global corporations, as the problem is likely to happen in other departments at different locations. Having a standard for solving certain problems will allow other employees to learn and solve these problems quickly and consistently.

We can also take one additional step to Step 7, which is to learn from all the previous steps and come up with measures that prevent similar problems from ever happening again. Doing this step correctly eliminates the need for the standards created in Step 6 and possibly even eliminates the need for using this problem-solving model for the problem. This takes the performance to the next level, as it would typically exceed the expectations of clients and management by stopping problems before they occur.

Project War Room

The last tool to discuss is the Project War Room. If designed correctly, it creates an environment for project teams to collaborate and manage project actions. The War Room displays project goals, schedules and resources so the project team can see the status of the project. It is also a place for the team to hold process meetings and problem-solving meetings. There should be an area for issues so that team members can see the problems that they are facing and can contribute to solve them. The Project War Room is a great tool for managing Core projects so further details will be presented in Chapter 9.

In summary, the tools presented in this section can be applied to manage project information from ideation to retrospective throughout the project lifecycle. I only provided a high-level description of these tools since I am afraid that too much detail will confine users when applying them. I want to emphasize that tools are the means to help us do things better. Defining the processes at a very detailed level will make implementation complicated and may lead to bureaucracy in administrating the use of tools. Using tools should not be a burden. It is up to management to introduce the right tools for the right situations and modify as they see fit. Many tools can be combined and they often overlap across different phases of the project life cycle model.

The next chapter will focus on managing tasks. Traditional Project Management uses the Work Breakdown Structure (WBS), which is a tool that does not promote innovation. I am proposing a new concept called Work Buildup Structure, a tool that incorporates innovation into the task definition process and helps project teams to exceed customer and market expectations.

Information Creation and Management

Exercise Questions

E-mail your thoughts on the questions to MBPM.Innovation@gmail.com
I will share my thoughts and answers to the questions.

Rules: (For details and reasons, please read Preface Page xvii)

1) One question at a time and state the Question # in the subject line of the email.

2) Provide a scanned copy of the book purchase receipt the first time you use an email to send in a question. This won't be necessary for future questions using the same email.

3) My response will be sent to you between 2-4 weeks after I receive your email.

Q5-1. Does the organization of this book make sense to you? Does it serve as a good example of how information can be presented? Do you have any suggestion for improvement?

Q5-2. The effectiveness of tools like Brainstorming, Brainwriting and Brainswarming are limited by the size of the team. For large-scale projects, what would you do to obtain good quality information in a reasonable time while making all members feel that their voices are being heard?

Q5-3. There is an exception for every rule, so do you see any types of meetings that do not fit the categories described in this book? If there are such meetings, how do you manage them effectively?

Q5-4. Micromessaging is noticeable through face-to-face interactions. Do you sense micromessaging in email? If so, what are the microinequities in email, and what are the microadvantages in email?

Q5-5. There are many theories and models in decision-making. Do you practice one that you feel is effective? How do you incorporate the effectiveness of the model that you use with the Six-step Decision-making model presented in this book?

Q5-6. There are also many problem-solving models. What are the major differences between the Convergent Problem-solving model and the other models that you use?

Chapter 6

The New WBS for Task Management

The New First Step in Project Planning

Under current mainstream project management, project planning starts with developing the Work Breakdown Structure (WBS), which is essentially a to-do task list for a project. From the top-level project objective, work elements are broken down level by level so that each activity can be clearly planned, owned, executed and monitored. WBS also serves as the foundation for project scheduling, budgeting, risk assessment and execution monitoring and control.

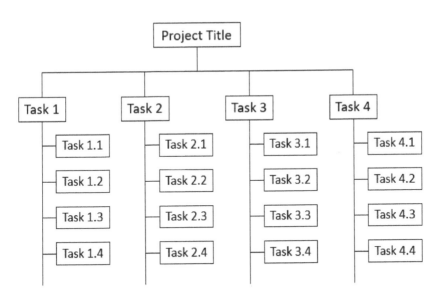

Work Breakdown Structure (WBS)

This popular project management practice, however, does not promote innovation. WBS is a top-down approach where tasks are developed from the top then broken down into further details. Once a level is defined, people naturally think within that level and generate more detailed tasks under it. Hence, it is difficult for people to think outside the box and create tasks that are not related to that level.

I am introducing a new approach, the Work Buildup Structure, which can also be abbreviated as WBS. To distinguish them, the new WBS is designated as WBuS. It incorporates innovation as defined in Chapter One. Instead of starting with the project title in a typical WBS, the project vision is used at the top level, which can be modified as innovative tasks are being developed. Tasks are generated from the bottom and building up to reach the vision.

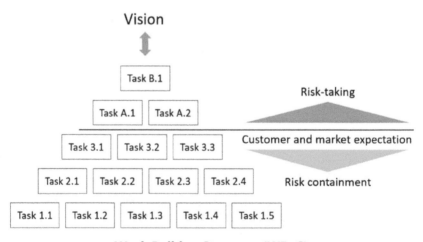

Work Buildup Structure (WBuS)

The WBuS approach demands the project team to interact with the customer and market to identify their expectations. Tasks that are within customer and market expectations are activities that must be performed well with minimal risks. Tasks that are above customer and market expectations are aiming at giving surprises to the customer and market. These tasks should be kept confidential, and it is perfectly okay that some of these tasks do not yield successful results. It would not hurt the reputation of the team or the company since no one outside the team knows about these failed tasks. Therefore, a psychologically safe environment is created to encourage risk-taking.

The "M" section of this chapter will explain the benefits of WBuS versus the traditional WBS through an example of building a house. The "L" section of this chapter will provide the specifics on how to develop WBuS for project task management. Once the tasks are generated, we can then shift our attention to project schedule development and management (Chapter 7).

Chapter 6

The New WBS for Task Management

Breakdown versus Buildup

Work Breakdown Structure (WBS) is a common project management tool, which serves the following purposes:

- Divide the overall project deliverables into distinct work elements
- Convert project requirements into manageable tasks
- Translate tasks into specific work packages for the team members
- Communicate the project scope to stakeholders to make sure nothing is missing
- Form the foundation for planning and monitoring

The most common way to perform WBS is by using a hierarchical structure starting with the project title or main objective. Then each level of the structure breaks the project deliverables down to more detailed elements and eventually down to the specific tasks. These specific tasks may then be assigned to project team members, enabling better estimations of cost, risk and time.

There are many ways of breaking down work packages, which can be product-oriented, function-oriented, task-oriented or resource-oriented. Product-oriented is breaking down the work based on the product's components such as the bill of materials (BOM). Function-oriented breakdown is based on the functions of the product such as electrical, mechanical, software, etc. Task-oriented is based on the process steps of the project such as design, prototype, test, etc. Resource-oriented is dividing the work based on the resources needed for those tasks such as engineering, legal, purchasing, marketing, etc. Combinations of these methods can be used and different methods may be applied at different levels.

Using the building of a house, as an example, WBS may look like the figure below. The first level breakdown is using a combination of task-oriented and function-oriented approaches. The second level WBS, under Plan and Foundation, are done by the task-oriented approach and the others are done by the function-oriented and product-oriented approaches.

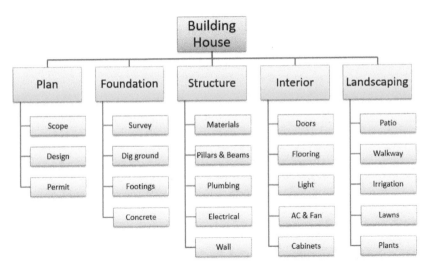

Project Work Breakdown Structure (WBS) Example

This top-down approach starts with the top level and typically continues by breaking down the tasks until each task can be properly managed. Once the WBS is developed, tasks are then scheduled, budgeted and monitored. Under this traditional approach, the project scope is clearly defined from the top and reflected on the WBS activities.

There are benefits to having the scope clearly defined but this confines the project team and its members to develop activities within the scope, making it difficult for the team to achieve results that exceed expectations. This approach does not promote innovation because there are no clear connections between activities and innovation. If management asks the project team to innovate, the WBS task owners will have difficulty determining what to do. In the house-building example, people do not associate most of the WBS tasks with innovation. How does the person applying for plan permit innovate? Same question goes to the person doing the foundation survey, the electrician doing electrical work, or the person installing the doors. There is no clear direction or coordinated effort to drive innovation in the project.

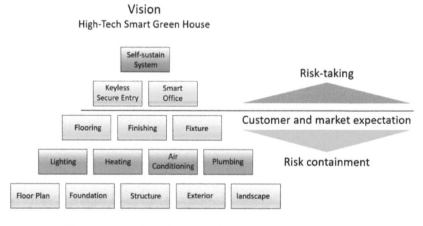

The New Project Work Buildup Structure (WBuS)

The new Work Buildup Structure (WBuS) proposed by this book can be used to design innovation into project task development. The figure above demonstrates the application of WBuS using the same example project of building a house. With WBuS, tasks are generated from the bottom starting from the basic features that everyone knows and expects from this type of projects, then builds the foundation level by level to reach the customer or market expectation line. Typically, tasks below the customer and market expectation line must be done well or the customer will not be happy and perhaps unwilling to accept or pay for the project. However, if we want innovation, which is defined as exceeding customer and market expectations, we need to continue to build on top of the expected tasks and generate tasks that could lead to surprisingly good results.

The concept of WBuS originated from my work in career advising. Employees who are just meeting their job requirements and expectations will most likely not be promoted. To help my advisees advance their careers, I use a similar model to help them identify activities that are below and above management expectations, and then develop tactics to enable them to perform efficiently at the low-level tasks while bring surprising results to their management with the high-level activities.

In the case of helping a person find a better job, we first identify things that are expected by the hiring manager through studying the job description and the hiring company thoroughly. But to increase the chance of being hired, the job seeker must identify things that are beyond the expectation of the interviewers and then find ways to promote and maximize the effects of those things, seeking to surprise the interviewers in order to deliver a good and long-lasting impression.

The key characteristics of WBuS, which are not present in traditional WBS are: 1) the team must interact with customers and understand the market to identify their expectations, 2) build a vision as well as tasks that lead to exceeding those expectations and 3) distinguish risk containment and risk-taking tactics based on tasks. The project team must manage risks associated with tasks that are within expectations, as failing these tasks would cause dissatisfaction from customers and the market. The tasks that are above the customer expectation line should be kept secret from the customer so that their results will be surprising. Keeping these tasks secret also provides a psychologically safe environment for the team to take risks; even if some tasks fail, there would be no impact to the customer as no one outside of the project team is aware that these tasks exist.

We also need to develop a vision of the project to drive the development of features and functions that are uniquely distinguished from existing solutions, which customers and the market already expect to receive. The vision must be inspiring so that the team is proud to be associated with the project and motivated to achieve it. Building these tasks and the vision is an iterative process that could take many attempts to refine. Sometimes the vision may be set too high and the tasks are too risky to achieve. The team should use the information management tools described in the previous chapter to generate, collaborate and decide on the best achievable vision along with tasks that lead to good surprises when presenting the project results to customers and the market.

Continue to read the "L" section of this chapter for the process of developing WBuS or go to the next chapter for building and managing the project schedules based on tasks developed through WBuS.

The New WBS for Task Management

Developing the Work Buildup Structure

Applying WBuS requires a new mindset. Traditional WBS is developed from a project owner's perspective. Typically, the question in the WBS developer's mind is what needs to be done to complete the project? WBuS is developed from a customer's perspective. It begins by the project team holding an ideation generating section using tools introduced in the previous chapter. The team may use Post-it notes to generate features and functions guided by the two questions shown in the illustration below.

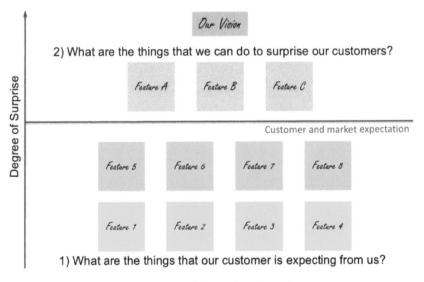

WBuS Feature and Function Development

As a bottom-up approach, the team should start with Question 1 in the bottom part of the illustration. What are the things that the customer is expecting from this project? While coming up with features and functions, keep in mind that if these items are not done well, the customer may not even want to pay for the project. To ensure this part is done with customer satisfaction, the team should have already interfaced with customers and have done some market research so that customer and market expectations are known. If there are uncertainties about some of the features and functions proposed, the team needs to seek more information.

Different color Post-it notes may be used to distinguish the degree of customer expectation with the strongest demands at the bottom. In the illustration, notes 1-4 on the bottom row represent features that the customer demands with certainty. Notes 5-8 on the second row from the bottom represent features that the customer desires with some degrees of freedom and flexibility. There could be more than two rows based on the degree of customer expectation. This process requires a good understanding of customer and market expectations and often cannot be done in just one session. Sometimes the degree of expectation is uncertain when a feature is written; it can be rearranged, however, when more information is available. Again, this is an iterative process. This bottom-up structure allows the team to understand the priority of the work, and generally the items at the bottom have higher priority than the ones higher up.

The vertical axis in the illustration is marked as "degree of surprise" instead of "degree of expectation" because "expectation" only applies on the features and functions that are below the expectation line. Features and functions above the expectation line are thus not expected by the customer so "surprise" is a better term. The two terms are inversely correlated as a higher degree of expectation means a lower degree of surprise.

When interfacing with customers and doing market research, the focus should not be on the needs of the customers but instead the problems that they are excepting to solve. There is a widely cited quote by Henry Ford: "If I had asked people what they wanted, they would have said faster horses." People cited this quote to support the idea that innovation is not usually generated from customer feedback but rather a few gifted visionaries who see what the customers do not see. People who hold this philosophy often ignore customer feedback. This book defines innovation as exceeding the customer and market expectations, so asking what the customers need and doing projects to meet their needs will not exceed their expectations. On the other hand, I also disagree that innovation can only be from a few gifted visionaries; interfacing with customers is still required but the intention is not to find out their specific needs but rather their expectations. Then, we identify features that go beyond their expectations to give them good surprises. That takes us to work on the top part of the illustration through generating features with Question 2 in mind: what are the things that we can do to surprise our customers?

The features A through C generated in the top part of illustration should result in great customer satisfaction, which means the customer is not only willing to pay for the project but will do so with a smile. The features generated in the beginning may constitute a high-level wish list with items that are typically broad and aiming in various directions. They may also require advanced technologies and significant resources to achieve. From the lists, a single vision of the project needs to be built, and based on that vision, the list of the features will need to be trimmed to develop a more focused list with achievable features. Therefore, finalizing these features is also an iterative process and may require several cycles to complete.

As mentioned earlier, these features should be kept secret to people outside of the project team since the objective is to give the greatest possible surprise to customers and the market. The team is encouraged to take risks and naturally it is perfectly fine if some of these secret features are dropped before the end of the project. As such, this process is carried out through the entire project lifecycle and it is not just done in the beginning planning stage. The features should be designed, experimented and reviewed regularly. New ones could be added while existing ones may be dropped. As the list of the features changes, the vision can be updated as well. Depending on the duration of the project, customer and market expectations may change such that some features may not be surprising anymore. Constant refinement of the list of special features and overall project vision is often necessary.

Typically done in the first round of ideation session, generating the preliminary features and functions in both regions above and below the expectation line is followed by generating the tasks required to achieve these features. When generating the tasks, the bottom-up approach is also applied. Think about what tasks absolutely need to be done first, then which tasks that may then be done to improve the quality or efficiency. As a rule of thumb, the details should be at the level where each task can be assigned to a single owner to ensure accountability. Also, each task should be at the level where it can be executed continuously for scheduling consideration, which will be further discussed in the next chapter when task schedules are built. Essentially, if there is a task that requires the owner to do certain things and then wait for a period before doing additional things for closure, it would be best to separate this task into two.

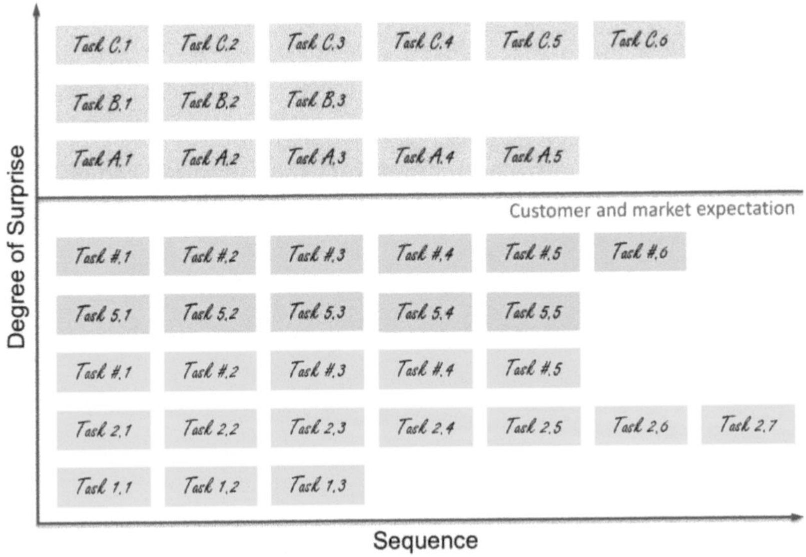

WBuS Task Development

As shown in the above illustration, tasks are generated based on individual features first. A task identifier should then be given to each task. In the illustration, for easy recognition, numbers are used for tasks below the customer and market expectation line and letters are used for tasks above the line. It is fine to have duplicate tasks as some features may require similar actions. Consolidation of tasks will be done in the next step.

For demonstration purposes, not all features in the illustration are shown and are represented by the "#" notation. While generating tasks, it is recommended to use correlated color Post-it notes as the ones used in the feature generation. Tasks are posted from left to right based on the sequence of actions. Don't worry about tasks that can be done in parallel; just put them down based on the preferred sequence. It will be addressed in future steps and when the schedule is being developed in the next chapter.

The next step is the consolidation of tasks across all features. This can be done by putting all the Post-it notes on a new board or in the same board by erasing the customer and market expectation line and text. Remember to take a picture for the record first. Afterwards, move the Post-it notes with similar tasks together with the later ones joining the earlier ones so the sequence still holds. Then refine the sequence of all the tasks and task groups to eliminate any conflicts.

Sequence

WBuS Task Consolidation

At this time, it is fine to have tasks or task groups to be done in parallel and some tasks may be done by different owners. Ownership assignment therefore needs to be done before the task sequence is finalized. As shown in the next illustration, ownership may be represented by color stickers with each color representing a different owner.

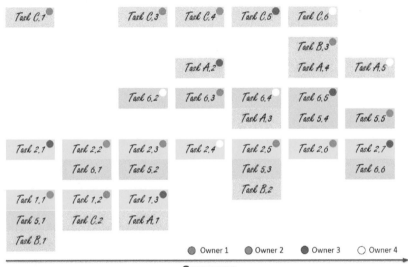

WBuS Task Ownership Assignment

This process should be done by the project team with all members present to determine the right owner for each task or task group. The intention is to apply the cultural approach to management, which was discussed in previous chapters. By working task assignments together as a group, the workload of each team member is shown openly, creating peer pressure on those who have less workload to either take on more tasks or support the team by taking on an additional share of logistical or administrative tasks such as running meetings and writing reports. It also ensures each task has a clear ownership with a firm commitment and agreement.

When tasks were initially generated, the guideline was for tasks to be separated to a level where a clear owner could be assigned for each. However, due to combining tasks into groups and changes made to some tasks to develop clear sequences as described in the last step, some tasks or task groups may now need to be executed by two or more owners. These tasks and task groups will now need to be broken down again so that a single owner can be assigned. Newly separated tasks or tasks groups could use a designation with a letter after the existing identifier such as "Task1.3a" and "Task1.3b" from the original "Task1.3." As such, there is no need to change the identifier designations of the subsequent tasks.

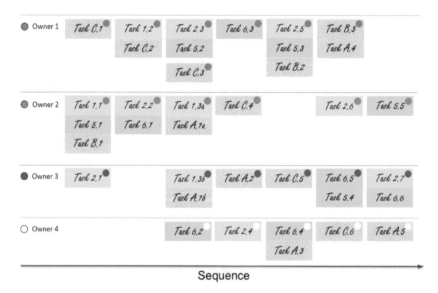

WBuS Preliminary Task Ownership and Sequence Map

After all the tasks and task groups are assigned with ownership stickers on the Post-it notes, they are arranged by owners as shown in the above illustration. While doing this, try to keep the tasks and task groups in the same column spaces so the overall task sequence remains intact. Notice that there may be multiple tasks or task groups in the same column owned by the same owner (e.g. Task2.3/5.2 and TaskC.3 assigned to Owner 1). A single person should not be doing two tasks at once. In the next chapter when scheduling is discussed, more explanation will be provided as to why multitasking should be avoided.

The multitasking situation needs to be addressed and there are several options for doing so. The first option is that Owner 1 still owns both tasks but prioritizes them by adjusting the sequence. In this case, Task 2.3/5.2 would most likely be the priority, as it is expected by the customer so is a minimum requirement. As a bottom-up approach, the lower level tasks in WBus are the foundation of the structure so these tasks are fundamental and actually have higher priority than the higher level ones.

The second option is to assign one of the tasks to another owner who has less workload, assuming the new owner has the skills and time to perform the task. The third option is to eliminate the task that is above the customer expectation line. Since such tasks are unknown to the customer, eliminating them will not directly impact the customer but will invariably reduce the innovativeness of the project. Dropping tasks should be the last resort and reserved for only when resources are highly constrained.

Based on the options selected by the team, the task sequence is further refined. The illustration in the next page shows the example of the first option, which is keeping both tasks with the same owner and prioritizing Task2.3/5.2 over TaskC.3.

Refined Task Ownership and Sequence Map

The task sequence map replaces the responsibility matrix used in Traditional Project Management. The responsibility matrix is usually a table with a vertical axis of tasks and horizontal axis for owners. The cell for a specific task and owner is typically marked with a letter designation of a role, such as R for responsible, A for accountable, C for consulted, and I for informed. In the task sequence map, however, it is not necessary to have roles on the tasks as each task owner is given the freedom and responsibility to pull resources from others to ensure that task is done satisfactorily. If an owner faces a problem with a particular task, he or she can initiate a problem-solving taskforce using the RAPID model discussed in Chapter 5. The roles of others could then be marked on the tasks using different types of stickers or designation systems. Models should be applied with flexibility but under the general direction of using culture to drive performance, which usually means involving the team and giving owners as much ownership as possible with fewer defined rules and less bureaucracy.

In the case of large-scale projects where teams are too big to have effective WBuS development sessions, the process can start with team leaders developing the high-level tasks that are owned by teams. Again, each task must have a single owning team. The team leads then bring the requirements back to their respective teams to develop the low-level tasks using the process described above.

Be flexible in applying the model, as even in large-scale projects there may be high-level tasks that are owned by individuals and not teams. Depending on the complexity of the task, some may require a team to handle them and some may be done simply by one individual. Therefore, high-level WBuS development meetings may not be manager-only meetings; we just need to keep the process as simple as possible and make sure the right people are participating in the right meetings.

Risk Handling in WBuS

The next step in task management is addressing task risks. This can only be done after clear ownership and sequences are defined. The task owners are the best candidates to perform risk assessments taking task priority into consideration. Many project management professionals believe that project management is essentially risk management. That is true for Traditional Project Management but if innovation is desired, risk-taking is necessary. If a company is not willing to take risks, it will most likely lose its competitiveness to others that are more willing to do so. Taking risk is easier said than done and just asking the project team and its members to take risks is not enough. Taking risks foolishly will also do more harm than good.

Traditional project risk management generally comprises the activities of risk identification, risk analysis, risk mitigation planning and risk monitoring and control. A typical Risk Profile used in Traditional Project Management is shown in the illustration below. Risk items can be plotted on the Risk Profile. High risk (HR) items have a high probability of occurrence and a high severity of impact. More risk containment efforts should be devoted to HR items. The mitigation tactics simply try to lower the risk probability, reduce the impact severity or both.

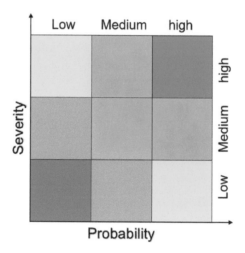

Typical Risk Profile in Traditional Project Risk Management

One of the major advantages of using WBuS versus traditional WBS is the risk-handling methodology which incorporates both risk containment and risk-taking as shown in the illustration on the next page. When done properly, WBuS clearly indicates where to take risks and where to control risks. By building tasks from the bottom along the axis of degree of surprise, the project team and its members should be clear that the degree of risk increases from the top to the bottom while the willingness to take risk increases from the bottom to the top.

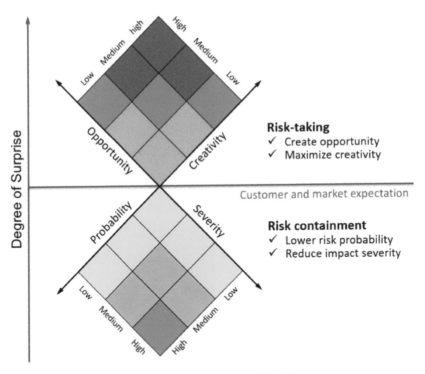

WBuS Risk-Handling Methodology

The color designations in the above illustration can be used to indicate task risk levels. Use different shapes of color stickers to indicate risk levels on the task Post-it notes in the Task Ownership and Sequence Map, or simply write acronyms on the task notes, such as HR (high risk, red), MR (med. risk, orange), LR (low risk, yellow), LS (low surprise, light green), MS (med. surprise, green) and HS (high surprise, dark green).

Traditional risk management is still applied for managing the risks of the tasks that are below the customer expectation line. For tasks that are above the line, risk-taking should be highly encouraged. What does risk-taking mean? It generally means maximizing creativeness and producing opportunities. Creativity is along the same axis of probability. Reducing the probability of risk typically involves taking a more conservative approach to reducing the chance of failure. As reliability theory indicates, the reliability of a system is the multiplication of the reliabilities of all its components. As such, a simpler system is more reliable since the probability of failure is lower, because there are fewer components. Maximizing creativity generally means encouraging the team to generate new features and functions, which will naturally increase the probability of failure. Since we are keeping these features and functions secret from the customer and market, the penalty of failure is much lower. On the other hand, if they are implemented successfully, the project will attract greater attention from the market and potentially have greater return.

Along the same axis as severity and above the customer expectation line is where you will find the opportunities to provide users with surprising features and functions. Task owners should develop use cases that allow users to see how these new features and functions could benefit them. If users do not know how to use the new features and functions, they will not know what to expect. These features and functions will therefore have no connection to the customer and market expectations and thus developing them would be a waste of time and effort. It is fine to generate more features and functions in the early stages of development with the understanding that not all of them will be successful. Following the "fail fast, fail forward" mentality, encourage project team members to test and modify the existing features and add new ones when they have slack time in the project. More focus and effort should be given to tasks that have a high surprise (HS) rating.

There is one type of risk that has not yet been mentioned: project schedule risk, which exists due to missing task due dates. If we use a scheduling approach that does not use due dates, there will be a new perspective on how we view and manage the scheduling risk. In the next chapter, Chapter 7, this scheduling approach will be introduced and tactics on managing task execution time will be elaborated.

Chapter 6

The New WBS for Task Management

Exercise Questions

E-mail your thoughts on the questions to MBPM.Innovation@gmail.com
I will share my thoughts and answers to the questions.

Rules: (For details and reasons, please read Preface Page xvii)

1) One question at a time and state the Question # in the subject line of the email.

2) Provide a scanned copy of the book purchase receipt the first time you use an email to send in a question. This won't be necessary for future questions using the same email.

3) My response will be sent to you between 2-4 weeks after I receive your email.

Q6-1. If a project team has difficulty in coming up with features and functions that are above the customer and market expectation line, what should the team do?

Q6-2. If a project team generates too many features and functions above the customer and market expectation line, and it is certain that there are not enough resources to try all of them, what recommendations do you have for the team?

Q6-3. The features and functions above the customer expectation line should be subjected to more changes than those below the line.

How do you propose to review and incorporate the changes? Should a change control system, like what is used in Traditional Project Management, be implemented?

Q6-4. While generating tasks, you will find that not all tasks carry the same degree of importance to the project. Do you think the important tasks should be highlighted? If so, how should it be done and what are the pros and cons?

Q6-5. We assign ownership to project tasks; however, some tasks may require resources other than people, such as facilities, tools, materials, etc. Should we incorporate the use of these resources into the task plan? If so, what should we do?

Q6-6. Task risks change as the project carries on, especially for those tasks above the customer and market expectation line. How often should we review and update the risk assessment?

Chapter 7

Scheduling for Better Time Management

Managing Schedules without Managing Due Dates

Traditional project schedules are developed based on task due dates, and schedule management means ensuring tasks are done by those due dates. Managing to meet due dates, however, rarely encourages acceleration of a project schedule.

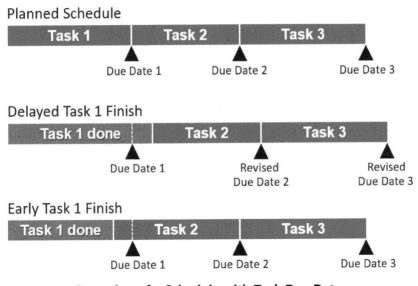

Execution of a Schedule with Task Due Dates

As demonstrated by the simple example above, if Task 1 is delayed during execution, the project manager will most like revise the schedule of the subsequent tasks to later due dates. If, however, for whatever reasons Task 1 is finished early, most project managers will not pull-in the overall project schedule; they usually keep the gained time as a safety buffer for possible delays in future activities. If these tasks are performed by different people, there is minimal incentive for the Task 1 owner to finish early, knowing that the overall project schedule will stay the same. As a result, the schedule is either on time or late; finishing early is very rare. This exemplifies the common observation known as the Parkinson's Law, which states that, "work expands so as to fill the time available for its completion" [33].

In 1997, Eliyahu Goldratt proposed the Critical Chain method in his book *Critical Chain* [34]. The major difference between Critical Chain Project Management (CCPM) and Traditional Project Management is how project tasks are scheduled; task due dates are eliminated by using buffers while the overall project targeted deadline remains. The concept can be viewed as in a relay race where each person does his or her best to complete the work and then pass along to the next person.

Two decades later, this scheduling method has not yet been widely practiced and I believe there are two main reasons for that. First, most people, especially managers, are accustomed to working towards deadlines. Managers are uncomfortable with team members working at their own pace, and they do not believe that employees will give 100% effort by just being asked. As demonstrated in the hand-raising exercise in Chapter 4 (Page L4-14 to L4-15), it is normal for people to not put in their best effort when first asked. Cultural and behavioral changes are therefore prerequisites for this method to work well.

The second reason is that most people practice this concept incorrectly by focusing on the scheduling part. The schedule using Critical Chain method is shown on the next page using the same earlier example. The only difference is the buffer gained by shortening all task durations. In many companies, managers constantly push their employees to work harder, which are often done by pulling in the deadlines in the traditional project scheduling method. While creating the schedule, most people do not see any obvious benefit to introducing a buffer. Naturally, they stay with the traditional scheduling method and use other means to shorten task durations if required.

Planned Schedule

Task 1	Task 2	Task 3	Buffer

Delayed Task 1 Finish

Task 1 done	Task 2	Task 3	Buffer

Early Task 1 Finish

Task 1 done	Task 2	Task 3	Buffer

Critical Chain Schedule without Task Due Dates

Through practicing the Critical Chain method, I have found that the method works well in most cases. The key to success is not in how the scheduling is developed but rather in how the schedule is managed in alignment with the cultural approach to management. To gain the most benefit, we need to understand the fundamental logic of using the Critical Chain method, which will be discussed in the next section. To understand how the Critical Chain schedule is developed and managed, read the "L" section of this chapter.

Starting from the next chapter, the focus will shift to the specifics of managing different types of projects, from PF projects to Core projects and then to CI projects.

Chapter 7

Scheduling for Better Time Management

Eliminating Bad Behaviors in Time Management

The Critical Chain method is relatively easy to understand and scheduling tasks using this concept is no more difficult than traditional scheduling. We need to understand, however, the assumptions and context of the concept to get the optimal result, as referenced in the triple-loop learning theory described in Chapter 4 (Page L4-22).

The Critical Chain method aims to eliminate many of the bad behaviors in time management. The schedule itself, regardless of how great it is, is not the main driver for behavioral change as it is just a plan on a piece of paper or in a computer program. It is how we manage the schedule that influences people's behaviors. Of course, task scheduling is a necessary part of that management system. It serves as a baseline reference to enable us to take better actions, and the actions that we take should align with the intention of the Critical Chain method, which is to eliminate bad behaviors to accelerate project task completions.

The first bad behavior is procrastination, which is exemplified by the student syndrome. Throughout years of education, from grade school to college, most people are accustomed to homework and exam deadlines and thus develop the habit of pacing their effort until the deadline is near as indicated in the graph on the next page. Since students are not experts in the subject being studied, they do not have total control over the plan so missing deadlines occur. Such habits are carried into the workplace as the same mode of work is presented in traditional project schedules with task deadlines.

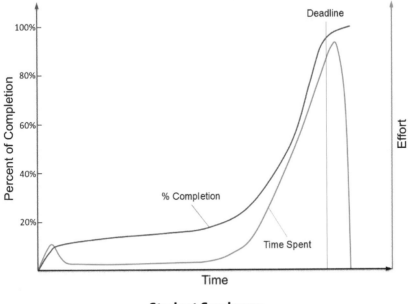

Student Syndrome

Pulling in a task deadline, which is the same as shortening the task duration, could reduce the tendency to procrastinate, but how could managers pull in deadlines without creating a very stressful environment for everyone? When the team anticipates that management will reduce task durations, they will typically put more padding into task duration estimates. This creates a tug-of-war between management and the project team, resulting in distrust in the work environment, which leads to the discussion of the second bad behavior in time management: padding.

When task owners are asked to estimate task durations, they tend to provide the estimated time between 70% and 90% certainty of completion. Of course, this also depends on the personality of task owners; some people are more conservative and only committed to what they can deliver, while others are more optimistic and willing to take risks. Most of them would not be comfortable to using the 50% mean average, as there is still a 50-50 chance of missing the deadline, and people typically want to have a higher success rate than a coin-toss probability. Traditional project scheduling methods do not offer any mechanism to encourage risk-taking, so even for those who love to take risks, offering a shorter time below the mean is rarely seen.

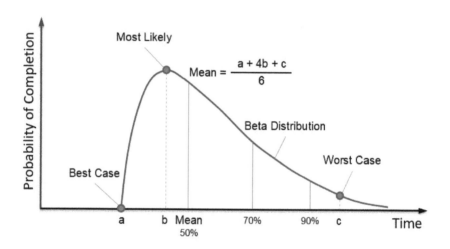

Task Duration Estimation

It is well known that the probability of a task completion is best represented by the Beta distribution as shown in the graph above. Time "a" is the best case scenario for completing a task, whereas time "b" is the most likely completion time and time "c" is the worst case scenario when most things go wrong. Traditional Project Management recommends the mean value to be used as the task duration to build the schedule. The mean can be calculated as (a+4b+c)/6, where there is a 50% chance for the task to be completed.

Instead of asking for single estimate from task owners, traditional project scheduling, such as Program Evaluation and Review Technique (PERT) uses three time estimates: optimistic, most likely and pessimistic to obtain the mean task duration. This reduces the task completion certainty from the 70-90% range to about 50%, but there is still a padding between the mean and the Most Likely duration. Looking from another angle, a task is likely to be completed sooner 50% of the time.

People also put padding when providing the Best Case, Most Likely and Worse Case estimates. If management is reducing the estimates that task owners provide, the task owners will put more padding into future estimates. As long as there are deadlines for tasks and one's performance is measured by meeting deadlines, people will put padding into the estimates to ensure their success.

The cause for padding is due to the concern of risks or interruptions. Interruptions are often a result of multitasking, which is the third bad practice in time management. Many managers believe that employees should develop the ability to handle multiple tasks, so it is common to see multitasking as a requirement in job descriptions. It is fine to require employees to learn and cover multiple areas, but that does not mean multitasking. True multitasking rarely happens, as most people cannot do two things at once. A typical task execution scenario with and without multitasking is shown on the next page.

Multitasking means switching from task to task, and switching from one task to another is basically an interruption. While the total time required for both tasks remains the same in both cases, the time required to finish Task 1 is extended with multitasking. What really matters is when a task is completed. Therefore, it is obvious that focusing on finishing one task at a time yields a better result.

With multitasking

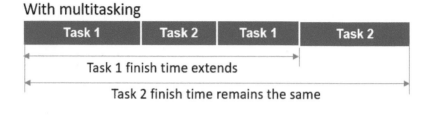

Without multitasking

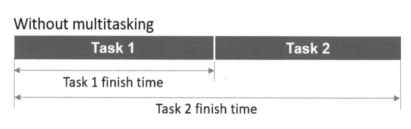

Task Execution with and without Multitasking

The illustration does not account for the inefficiency of switching between tasks. In reality, when we are switching from one task to another, it typically requires a ramping period for us to get back into focus, so the situation is worse than shown. A similar situation occurs in production where equipment is dedicated to specific products as much as possible to minimize setup and switching time. In manufacturing, production schedules are generally more succinct, compressed and visual compared to project schedules. The Lean philosophy with continuous waste elimination contributes to these characteristics of production schedules. Many of the Lean practices can be applied to project management.

People who have implemented Lean will say that Lean is a cultural shift and requires behavioral changes. As discussed in Chapter 4, the focus of a corporate culture is the will of employees. To achieve better time management in project execution, project team members must willingly change the bad behaviors in time management, namely reducing procrastination, padding and multitasking. Culture is built on the corporate system; a good time management system is therefore required to enable that culture. The next section will present a project time management system built on the Critical Chain method, and it utilizes the cultural approach to manage schedule performance.

Scheduling for Better Time Management

Developing and Managing Project Schedules

The first step in developing a project schedule is estimating the duration of each task generated from WBS or WBuS, which can be either bottom-up or top-down. The bottom-up approach involves asking task owners to provide estimates. As mentioned previously, people have the tendency to put padding into their estimates. For better accuracy, project managers can request three estimates, Best Case, Most Likely and Worst Case, and then using the formula $(a+4b+c)/6$ to get the average duration. The top-down approach assigns the duration to task owners based on historical data, expert opinions or timelines set by management based on business needs. Regardless which approach is used, an estimated duration associated with each task is needed before we start scheduling.

Using these estimated task durations, a project schedule can then be built using Gantt charts and network diagrams. There are many network scheduling tools used in Traditional Project Management, such as Activity-on-Note (AON) and Activity-on Arrow (AOA) diagrams, the Precedence Diagram Method (PDM), the Program Evaluation and Review Technique (PERT) and the Critical Path Method (CPM). All of these tools use due dates for tasks with emphasize on the importance of achieving those due dates as the main tactic in managing a schedule. As discussed earlier, scheduling based on task due dates precipitates many bad behaviors in time management. The Critical Chain scheduling method, along with the execution management practices introduced by this book, will eliminate many bad behaviors for better time management.

To demonstrate how a project schedule is developed and managed under the Critical Chain method, let's use the following example.

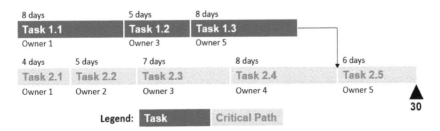

An Example Schedule in Traditional Project Scheduling

Under traditional project scheduling, it is clear that tasks with 2.# are on the critical path of the project. Any delay in one of these tasks will push the project schedule beyond the 30-day target, unless one or more of these tasks are shortened to compensate for the delay. In addition, the schedule shows that Owner 1 and Owner 3 must multitask to meet the deadline of the project. There is not enough slack time (3 days) on the Task1.# chain to allow either Owner 1 or Owner 3 to work one task at the time. If these are bottom-up estimates, the task owners may have already included padding for multitasking in their estimates, as they must to avoid putting themselves in a tight situation to meet the deadlines.

Let's now build a Critical Chain schedule based on this example. The first step is to reduce the original time estimate by a certain percentage; fifty percent (50%) is recommended. Please do not jump to the conclusion that the work time is cut significantly and that this is a means to simply force everyone to work harder. This 50% shortened time is not the allocated work time for the task owners. This step can be done simply by dividing the original tasks as shown below.

First Step in Critical Chain Scheduling

The next step is to realign the tasks based on the new shortened durations, starting with the first task on the original critical path, which is Task2.1. Then push other tasks to avoid multitasking. If both tasks are starting at the same time, such as Task1.1 and Task2.1 by Owner 1, the task on the original critical path takes priority. If not, the task that starts earlier typically takes the priority and the task that starts later will be pushed to begin when the first task ends (e.g. Task1.2 by owner 3 is pushed to start after Task2.3). Task1.3 and Task2.5 by Owner 5 is another example with Task1.3 starting first even though Task2.5 is on the original critical path. The new schedule is shown in the diagram below.

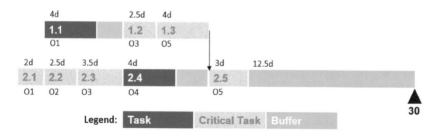

Second Step in Critical Chain Scheduling

This schedule has a new critical path because it is now people-oriented versus task-oriented as in the original schedule. The new concentrated critical path has a total time of 17.5 days, which leaves a 12.5-day period to the target project completion. According to Goldratt, this period is called the project buffer. Buffers for non-critical path tasks, such as Task1.1 and Task2.4, are called feeding buffers [34]. Others who practice Critical Chain scheduling also categorize additional buffers such as milestone buffer and resource buffer [35]. These buffer distinctions help to understand the schedule but they make no difference in managing the schedule execution. As mentioned earlier, the main benefit of this approach is not from how the schedule is developed but from how the schedule is managed. Therefore, I just use the term buffer for all and keep it simple.

Most project managers who practice the Critical Chain scheduling method stop here and consider that the schedule is done. They move on to managing execution as if the due dates are shortened with a buffer created. Obviously, most project team members and task owners would not welcome this method with open arms. While it reduces multitasking for certain owners, the combined task work time is not much different from the original estimate. In the example, while Owner 1 had a maximum of 8 days for multitasking both Task1.1 and Task2.1, it is now 6 days. For Owner 3, the maximum of 7 days for both Task1.2 and Task2.3 is now also 6 days. For those owners who do not multitask, however, such as Owner 2 and Owner 4, the work time is significantly shorter. As mentioned in the earlier section, this is one of the reasons that Critical Chain scheduling, while originally proposed in the 1990s, has still not been widely accepted.

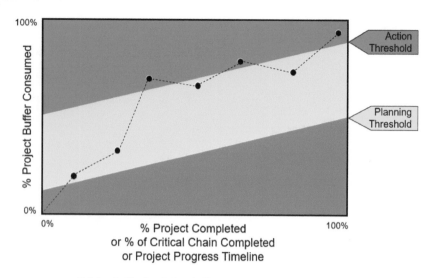

Critical Chain Scheduling Fever Chart Example

In the project execution phase, mainstream project managers who practicing Critical Chain scheduling use a tool called a "Fever Chart" to manage project buffer consumption [35]. A typical Fever Chart is shown above. Periodically, these project managers perform an assessment on the percent of the Critical Chain tasks completed, and the percent of buffer consumed, and then plot the point on a chart. It is called the Fever Chart because the chart has green, yellow and red regions. A dot in the red zone means that the buffer is consumed at a higher rate than the critical task completion, alerting project managers to take actions to reduce the buffer consumption rate.

This tool can be used as a high-level indicator for project status, but I found it difficult to use and the benefit is minimal. At the time of assessment, there are partially completed tasks that require project managers to determine the percent completion plus the estimated buffer consumption. Most experienced project managers would probably agree that assessments of partial completed tasks are generally difficult because of subjectivity; the last 20% of work could take 50% of the time. It also requires the task owners to provide honest data. To best utilize the Critical Chain method, there are better ways to manage the schedule but extra steps in planning are needed before moving on to execution management.

	Task2.1	Task2.2	Task1.1	Task2.3	Task1.2	Task2.4	Task1.3	Task2.5
Duration	4	5	8	7	5	8	8	6
Ideal Start	0	2	2	4.5	8	8	10.5	14.5
Target Start	0	3.4	3.4	7.7	13.7	13.7	18	24.9
Owner 1	2		4					
Buffer	1.4		2.9					
Owner 2		2.5						
Buffer		1.8						
Owner 3				3.5	2.5			
Buffer				2.5	1.8			
Owner 4						4		
Buffer						2.9		
Owner 5							4	3
Buffer							2.9	2.1

Buffer Allocation in Critical Chain Scheduling

The next step is to calculate the buffers and allocate them to each task. Shown above, a summary table using spreadsheet applications may be used to complete this step. This is done by giving the corresponding amount of buffer to each task based on the percent of that task over the total time. Below is the buffer calculation for a 3.5-day task from the example:

$$3.5\text{d Buffer} = \frac{3.5\text{d x } 12.5\text{d}}{17.5\text{d}} = 2.5\text{d}$$

This buffer calculation should be used for all tasks and not just the tasks on the critical path. Some non-critical tasks may have more buffer available, but it is simple to keep all buffer calculations the same.

Consistent buffer calculations also ensures fairness in managing the performance of all tasks owners, which will be further explained when discussing schedule execution management.

With the calculations done, we then communicate to task owners that the allocated work time of a given task is the shorten time plus the buffer. For Owner 1, the combined time for both Task1.1 and Task2.1 is now at 10.3 days; this is without multitasking, which is 29% more than the original 8 days with multitasking. Similarly, for Owner 3, the combined time for both Task1.2 and Task2.3 is also 10.3 days without multitasking, which is 47% more than the original 7 days with multitasking. It should now be easier to convince these owners to accept this method. Actually, they should have a very high probability of finishing these tasks early. For the owners who are not multitasking, Owner 2 now has 4.3 days instead of the original 5 days and Owner 4 has 6.9 days instead of 8 days. There is a small drop but it can be explained that we are targeting at the Most Likely time as seen in the Beta Distribution Graph (Page M7-3) as opposed to the average time.

The importance of this method is on how the schedule is managed during project execution. As mentioned before, we are not managing the schedule by task due dates, so it is fine if a task runs over the shortened time (e.g. Task2.1 takes over 2 days). If Task2.1 is done within the 3.4-day combined time with buffer, the project is still on schedule. In other words, if all tasks are finished within the combined time, the entire project will hit the targeted completion date. Even if a few tasks are done over the combined time, the overall project still has a good chance of finishing on schedule because some of the task owners are more likely to finish early if we manage the schedule correctly.

We manage the project schedule by managing buffer consumptions as opposed to due dates. The chart shown below is used to indicate the performance of each individual task and its owner. When a task is complete, we use the actual completion time minus the shortened time and then divide by the buffer time to get the percent of buffer consumed for that task. Essentially, we are measuring the performance of task owners based on how well they utilize the buffers.

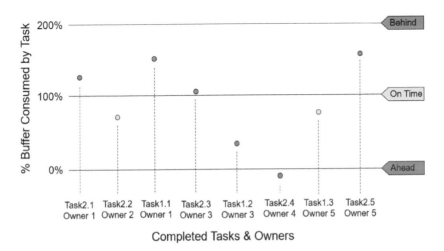

Individual Task Buffer Consumption Chart

In traditional project scheduling, if one of the tasks in the critical path runs over then the entire project is likely to be delayed, particularly if this delay occurs with an earlier task. For instance, if Task2.1 runs over, it creates a chain reaction to revise due dates of all the other Task2.#. This also gives a longer slack time to non-critical path tasks, which will most likely promote procrastination and lead to delays in non-critical path tasks as well. Once a delay occurs, it impacts the subsequent tasks and also provides an excuse for the following task owners to justify further delays.

In a typical project, there will be delays in some of the tasks. If all task deadlines are met, the schedule might be too relaxed; task owners are either not being challenged or there is too much padding in their estimates. On the other hand, if multiple tasks are delayed, managing performance of individuals becomes difficult. For instance, at the end of the project or during an annual performance cycle, if the manager attempts to take disciplinary action against Owner 5 based on the delay of Task2.5, Owner 5 can easily defend the delay by citing that there are delays in many tasks before Task2.5 and can challenge why he or she is the one who is being disciplined. When employee perceive unfair treatment, there will be distrust and managing the team going forward will be difficult.

The indicator chart of individual task buffer consumption solves this problem. Buffers are proportionally allocated based on the original estimated durations. Monitoring the percent of buffer consumed is therefore fair to all task owners. Longer tasks have longer buffers allocated. When there are many tasks experiencing delays, which may be perfectly fine as it indicates that employees are being challenged and stretched out of their comfort zones, the manager can still address the performance issue of Owner 5 by observing that Task2.5 has the highest percent of buffer consumption. We now can measure each individual's performance with an indicator down to decimal points. While not the primary purpose, this indicator allows management to manage performance based on data. The essence of this indicator is to enable the cultural approach to management that was discussed many times in previous chapters. How? It displays the individual performance in the context of a shared community.

The main buffer in Critical Chain scheduling belongs to everyone in the project. If someone uses more than their allocated buffer, meaning over 100%, they are using a community resource. Anyone who repeatedly uses more than their own share of the community buffer should be pressured by peers to improve performance. Once we educate team members on this notion and show the indicator in the Project War Room or a community space, everyone on the team can see everyone else's performance. Those who use more than their share of buffer are pressured to improve if they want to continue be part of the team. This peer pressure often drives those individuals to work overtime in an attempt to perform equally with the majority. If they do not try hard to improve, who would like to be in the same team with them again in the next project?

When a project schedule is developed, bottom-up duration estimates are preferred as team members are more likely to commit to the tasks when they are involved from the beginning. It is nevertheless human nature to provide padding in those estimates. As people understand the Critical Chain method, some will put enough padding into the original estimates to reduce the chance of using the community buffer. If an individual consistently uses less buffer and stays below the 100% line, the manager should have a performance discussion with this individual. As Intel founder Gordon Moore once said, "If everything you try works, you aren't trying hard enough." This is case where managers need to be very careful about rewarding individuals based on project results. Based on this indicator, the individual performs well but if he or she is rewarded, it actually encourage employees to put more padding into their estimates.

When I was teaching this method in China, many students told me that their managers do not use bottom-up estimates. They simply determine the Best Case time and build the schedules; they do not care about the Most Likely time nor the average time. In such top-down situations with extreme time pressure, this method is more beneficial as it promotes a fair working environment and enables incentives for accelerating task completions.

The first step in applying this system to the above case is still collecting bottom-up estimates. The next step is building the Critical Chain schedule as before. Of course, this schedule will not meet the aggressive project completion goal set by management. Now we set the top-down mandated project completion date as the end date of the schedule, and then do a percent reduction for all tasks and buffers. In the above example, the project ends in 30 days. If management demands that the project be completed in 24 days, we then reduce all tasks and buffers by 20%. This reduction is applied across the board so it is fair to everyone.

In addition, we can offer incentive rewards to team members who use less allocated buffer. For example, for every percentage point of buffer not used, the task owner is given a $100 reward. Again, we have to be careful for giving out rewards. To be fair for all, we need to offer everyone a chance to get it if they put in the effort. Therefore, this reward should be available for all task owners, not just the owners of critical path tasks. The reasoning is the same as for using a consistent buffer calculation for all tasks regardless of whether they are on a critical path or not. We have previously discussed that money is not a good motivator unless the work is mechanistic. As such, if we can specify the work mechanistically using clear and objective measurements, rewards and incentives would work better. In this case, rewards are most likely effective as they are given strictly based on a straightforward calculation of buffer saved.

The basic concept of Critical Chain scheduling and its execution management has been explained, but there are many details and variations in practice depending on the nature of a project. In the last chapter, an example was used to explain the steps of WBuS and we produced the last Task Ownership and Sequence Map as shown on Page L6-10. Continuing from the WBuS build, we use the same example to further demonstrate how a schedule is developed and managed, especially on the incorporation of tasks above customer and market expectations. The first step is to ask task owners to estimate the time for each task and combined tasks. The time estimates can be simply written on the existing Post-it notes in the Task Ownership and Sequence Map or with new, smaller Post-it notes on top of the existing ones. Based on the time estimates, the schedule is built as shown on the next page. Arrows are used to indicate the sequence of tasks.

At this juncture of the scheduling process, the tasks with letter designations, such as TaskC.1, TaskC.3, TaskA.2, TaskC.4, TaskC.5, TaskB.3/A.4 and TaskC.6, are not being considered. These tasks are above the customer and market expectation line, which are intended to produce surprising results to the customer and market and kept confidential within the team. Therefore, these tasks should not be on the critical path of the project. That is the reason that we build the schedule using the tasks with number designations first since they are required and have higher priority.

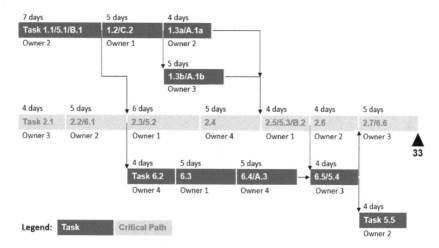

Schedule with Bottom-up Time Estimates

Next, the task durations are divided as shown below. Fifty-fifty is recommended but in some situations, different percentages may be experimented to adjust the buffer to a desirable size.

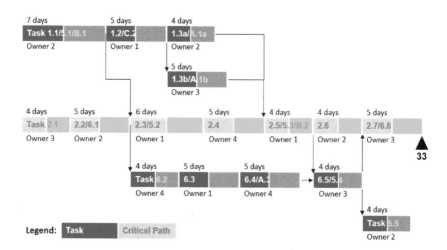

Schedule with Time Estimate Division

Afterward, the shortened tasks are realigned under two conditions: no multitasking and maintain task sequence. The first condition is no multitasking. Parallel tasks performed by the same owner will need to be rearranged. The order is earlier task first, and if both tasks start at the same time, the critical path task comes first. In the example, Task2.2/6.1 performed by Owner 2 is shifted after Task1.1/5.1/B.1. The second condition is keeping the task sequence (e.g. Task6.2 is after Task2.2/6.1 even if Owner 4 is free from the start and can perform a task). The Critical Chain schedule is shown below.

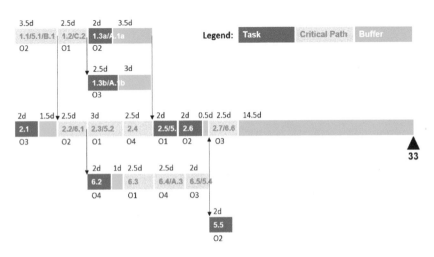

Critical Chain Schedule

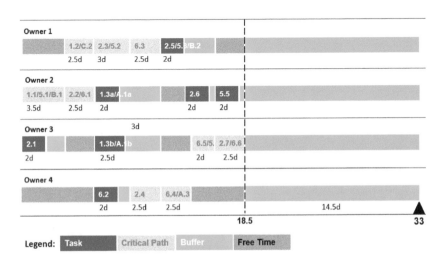

Critical Chain Schedule by Owner

The WBuS Task Ownership and Sequence Map (Page L6-10) is shown by owners. Hence, to provide a better view from a task owner's perspective, the schedule by owner diagram is also preferred. It is shown on the previous page under the Critical Chain Schedule.

In addition, a summary table is generated to provide easy viewing and execution management, which is sown below. Critical path tasks are indicated by red fonts.

Task	1.1 5.1 B.1	2.1	1.2 C.2	2.2 6.1	2.3 5.2	1.3a A.1a	1.3b A.1b	6.2	2.4	6.3	6.4 A.3	2.5 5.3 B.2	2.6	6.5 5.4	2.7 6.6	5.5
Duration	7	4	5	5	6	4	5	4	5	5	5	4	4	4	5	4
Ideal Start	0	0	3.5	3.5	6	6	6	6	9	9	11.5	11.5	13.5	14	16	16
Target Start	0	0	6.2	6.2	10.7	10.7	10.7	10.7	16.1	16.1	20.5	20.5	24.1	25.0	28.5	28.5
Actual Start																
Actual End																
Owner 1			2.5		3							2.5	2			
Buffer			2.0		2.4							2.0	1.6			
Actual																
% Used																
Owner 2		3.5		2.5		2								2		2
Buffer		2.7		2.0		1.6								1.6		1.6
Actual																
% Used																
Owner 3		2					2.5								2	2.5
Buffer		1.6					2.0								1.6	2.0
Actual																
% Used																
Owner 4								2	2.5		2.5					
Buffer								1.6	2.0		2.0					
Actual																
% Used																

Critical Chain Schedule Table View

Now, let's take care of the letter-designated tasks. Using the Critical Chain Schedule by owner diagram, the letter-designated tasks are added to the respective owners. For TaskC.1, Owner 1 has free time available. TaskC.3 is the most difficult to accommodate as it is on top of a critical task. The team needs to identify a solution, which may include assigning this task to another owner or reducing the scope of feature C. For TaskB.3/A.4, Owner 1 has free time at the end but not enough.

However, there is a 2.5-day buffer before the free time and Task2.5/5.3/B.2 only uses 1.6 days of buffer so it should be fine. The situation is similar for TaskC.4 by Owner 2 with the exception that TaskC.5 needs to be accommodated in Owner 3's free time first so TaskC.4 is pulled in. For TaskA.2 by Owner 3, the buffer after Task1.3b/A.1b may not be enough to cover. However, there is free time before Task1.3b/A.1b. Owner 3 should look for opportunities to use that free time for pre-work to ensure Task1.3b/A.1b is completed in the Best Case time. Owner 4 has no problem accommodating TaskC.6. TaskA.5 is pushing into the project buffer because TaskB.3/A.4 needs to finish first, but if Owner 1 finishes Task2.5/5.3/B.2 with less allocated buffer, TaskB.3/A.4 can start sooner so this issue is not too serious. The following illustration shows the inclusion of letter-designated tasks.

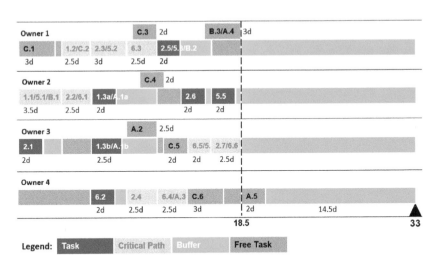

Final Critical Chain Schedule By Owner

With the schedule finalized, we are ready for execution management. First, we utilize the Critical Chain Schedule table to setup a status board available for everyone to see and update. A computer version in Excel with cell formulas can be set up to minimize data update effort. As shown below, the only update needed is the actual end time of each task upon its completion. Everything else is calculated.

Task	1.1 5.1 8.1	2.1	1.2 C.2	2.2 6.1	2.3 5.2	1.3a A.1a	1.3b A.1b	6.2	2.4	6.3	6.4	2.5 A.3 5.3 B.2	2.6	6.5 5.4	2.7 6.6	5.5
Duration	7	4	5	5	6	4	5	4	5	5	5	4	4	4	5	4
Ideal Start	0	0	3.5	3.5	6	6	6	6	9	9	11.5	11.5	13.5	14	16	16
Target Start	0	0	6.2	6.2	10.7	10.7	10.7	10.7	16.1	16.1	20.5	20.5	24.1	25.0	28.5	28.5
Actual Start	0.0	0.0	6.5	6.5	12.0	11.5	11.5	12.0	17.0	17.0	20.5	20.5	25.0	25.0	28.0	28.0
Actual End	6.5	4.0	11.5	12.0	17.0	15.5	16.0	16.0	20.0	20.5	25.0	25.0	28.0	28.0	32.0	32.0
Owner 1			2.5	3							2.5		2			
Buffer			2.0	2.4							2.0		1.6			
Actual			5.0	5.0							3.5		4.5			
% Used			128%	85%							51%		159%			
Owner 2	3.5			2.5	2									2		2
Buffer	2.7			2.0	1.6									1.6		1.6
Actual	6.5			4.5	4.0									3.0		4.0
% Used	109%			100%	128%									64%		128%
Owner 3		2					2.5							2	2.5	
Buffer		1.6					2.0							1.6	2.0	
Actual		4.0					4.5							3.0	4.0	
% Used		128%					102%							64%	77%	
Owner 4								2	2.5	2.5						
Buffer								1.6	2.0	2.0						
Actual								4.0	3.0	4.5						
% Used								128%	26%	102%						

	#	%		#	%		#	%		#	%
<50%	1	6.3%	50% - 99.9%	5	31.3%	100% - 150%	9	56.3%	>150%	1	6.3%

Critical Chain Schedule Tracking Table

The ideal start time is based on the shortened time of the task without any buffer, which is nearly impossible in most projects. We still display them as stretch goals for the team and task owner. The target start time is the combination of the shortened time and the allocated buffer, which assumes 100% of the buffer is used for each corresponding task. If the team is hitting all the target starts, the project finishes on time. This schedule tracking table can replace the Individual Task Buffer Consumption Chart shown on Page L7-8. The buffer consumption percentage is updated after each task finishes. Color codes are used to show the performance of each task. An interesting fact is that one task finishing with less allocated buffer is enough to hit the overall project schedule. As shown in the example, a majority of the tasks (62.6% = 56.3% + 6.3%) are finished using more than their allocated buffers but the project still finished one day early. Compared to the traditional project scheduling, one delay task may already be enough to delay the entire project.

At the end of project, this data can be used to evaluate the overall performance of each team member and provide fair and data driven compensations. As discussed before, for fairness in evaluating performance, the same buffer calculation is used for all tasks. Clearly, there are some non-critical path tasks, such as Task2.5/5.3/B.2, Task1.3a/A.1a, and Task1.3b/A.1b, which have longer buffers on the schedule diagram, but the table shows the calculated buffer size using the same formula. All task owners receive the same treatment. The data also enable performance comparisons among different projects. There are many indicators that can be generated, such as overall percent of buffer consumed, percent of tasks started within the target start time, percent of critical tasks over allocated buffers and percent of critical task buffer used versus non-critical task buffer used.

Since the table view does not show the trend of the project performance, we can track the project status using a trending chart as shown on the next page. It replaces the Fever Chart shown in Page L7-5, but it is not time-based. An update is done at the completion of each task and not in regular time intervals, which eliminates the need for assessing partially completed tasks. The calculation is simply done as the percent of buffer consumed minus the percent of task completed. Percent of task completion is calculated using task allocated time divided by the overall project time. The calculation formulas can be built in an Excel file using the data from the tracking table. Once the actual end time of a task is entered, the chart will be automatically updated. The chart below takes the data from the table on Page L7-17. (The Excel file is available for request through email if you purchase the book.)

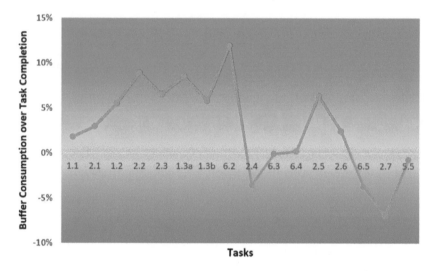

Buffer Consumption over Task Completion Trending Chart

Traditional project scheduling, including the current mainstream Critical Chain scheduling if practiced, is generally task-oriented with the intention to design a system around tasks. For instance, traditional scheduling starts with task placement without concern for task owners doing multiple tasks at the same time. The Fever Chart proposed in Critical Chain schedule monitoring and control is also focused on task performance only. The schedule management practices introduced in this chapter are oriented towards people: project team members. I am sure that there are some details missing when this method is applied to certain projects, but as I emphasize repeatedly, no single tool is suitable for all situations; modifications are necessary and better outcomes can be achieved if we take the people-oriented approach.

By now, the main project management strategies and tactics have been covered in all the chapters thus far. Those who are experienced in Traditional Project Management should find that the topic of cost management is absent, which is a big topic under Traditional Project Management. In my opinion, cost management is the most important focus in managing operations and productions but not in managing projects. Projects are for the future of a company and if the company is not willing to invest and invest adequately, it will fall behind its competitors. We do need to be cost conscious and use resources efficiently. Achieving efficiency in managing projects means effectively managing people, information, task and schedule, which are the areas that we just covered in the last four chapters.

Topics of cost management such as budgeting will be covered in the next three chapters corresponding to the management of different categories of projects. Planning and managing the budget of a Core project is very much different from planning and managing PF and CI projects. In the next chapter, we start to discuss the specifics of managing PF projects.

Chapter 7

Scheduling for Better Time Management

Exercise Questions

E-mail your thoughts on the questions to MBPM.Innovation@gmail.com I will share my thoughts and answers to the questions.

Rules: (For details and reasons, please read Preface Page xvii)

1) One question at a time and state the Question # in the subject line of the email.

2) Provide a scanned copy of the book purchase receipt the first time you use an email to send in a question. This won't be necessary for future questions using the same email.

3) My response will be sent to you between 2-4 weeks after I receive your email.

Q7-1. Is it possible to use incentives and rewards in the traditional project scheduling approaches to encourage project teams to finish projects early?

Q7-2. In many cases, multitasking is unavoidable, as excursions, customer requests and management priority changes are typical in most workplaces. What should managers and workers do to reduce the impact of multitasking?

Q7-3. Traditional project scheduling puts a lot of effort into defining task relations by using various networking diagrams and methods. The approach introduced in this book does not seem to spend much effort on task relations. Why and do you feel if something is missing?

Q7-4. We briefly mentioned using the Critical Chain schedule for tracking data to develop indicators for comparing the performance of different projects (Page L7-18). What indicators are best for project performance benchmarking?

Q7-5. Change will happen during project execution, such as new features and requirements from customers or management. How do we manage changes in this new approach to schedule management?

Q7-6. Is this new schedule management approach capable of handling large numbers of tasks and big project teams? If so, how can it be applied?

Managing Path-finding Projects

Vision from Passion

Path-finding (PF) projects are for the future as shown in the vision and mission space in the MBPM strategic chart below. Therefore, before starting to plan for PF projects, the vision and mission of the company need to be developed.

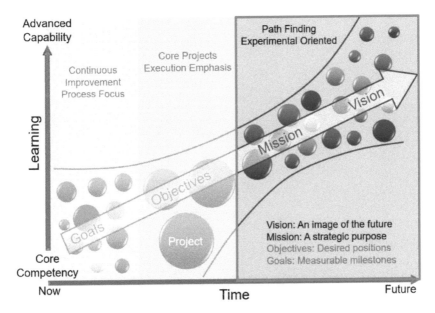

Path-finding Projects in MBPM

The vision is the aspirational picture of the future and the mission is the strategic purpose of the organization. A company's vision and mission serve two main purposes. The first is to provide a general common guide for all the employees in the company, like the North Star in the sky guiding people in the dark. The second is being the purpose motive that was discussed in Chapter 4 Page M4-3. The vision and mission are what employees in the company are proud to say to an outsider when asked about what they do.

How do we develop an inspiring vision that guides the company's path-finding effort and inspires employees to do great things? It starts with passion. When a company is founded, the founder typically develops the vision from his or her passion, and that vision attracts individuals with similar passions to join the company. The path to success for a company can be explained through the TOP model used in career development as shown below.

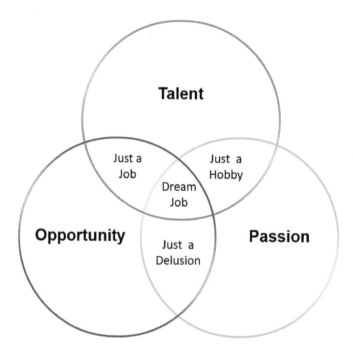

TOP Model in Career Development

In the TOP model, T stands for talent, O for Opportunity, and P for Passion. A person is most likely to be successful if he or she can land the "dream job," which means the person has the passion for the job, the talent to do it well and the job opportunity exists. Among the three elements, the passion circle is the most stable one. Since it is rather personal, it is where we start in planning a career for an individual. The talent circle increases as the individual develops and learns. More talent means a higher chance to find a dream job. The opportunity circle is the most unpredictable one as it can shift, shrink or expand.

Likewise, when applying the TOP model to an organization, it begins with a shared passion of the majority of the employees in the organization. The organization's vision and mission statements represent the passion that should be clearly conveyed to all employees. To stay aligned with changes in the environment and the employee base, an organization often renews its vision and mission. As I am writing this book, Stanford University is in the process of renewing its vision. The university created the website "ourvision.stanford.edu" and over 2,800 people contributed to the development of the new vision of Stanford: "Navigating a dynamic future - We will spark knowledge and creativity, advance learning, and accelerate impact for the benefit of humanity" [36]. The development of the new vision and mission help to realign the passion of the employees in the organization.

After establishing the passion circle, the focus then shifts to the talent circle, which is also under the control of the organization. Developing talent means obtaining capability by learning, which corresponds to the learning axis of the MBPM model. PF projects serve as stepping stones to gain the capabilities needed to achieve the vision. PF projects should therefore be defined based on the learning scale, from current core competencies to advanced capabilities.

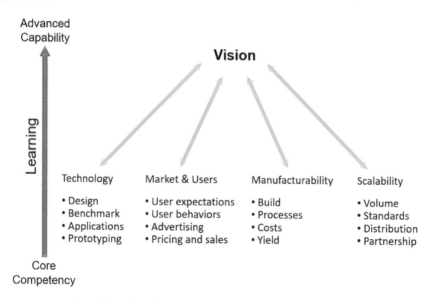

Defining Projects Based on the Learning Scale

As illustrated in the above figure, an organization should explore the portfolio of PF projects covering all the major areas that lead towards its vision. In today's high-tech world, technology development is a natural focus when developing PF projects. However, PF projects should not be just technology centric but include capabilities needed to take the technology to the market along with manufacturability and scalability. When an organization is capable in all major facets of the business it has a greater opportunity to succeed.

That takes us to the opportunity circle, which is largely determined by the development cycles of technology and market. Although the organization has the least amount of influence in this circle, understanding the trends of technology and market development will enhance an organization's ability to predict the future. In the M section of this chapter, we will discuss the topic of emerging technology and market management, which helps an organization to develop its capabilities. In the L section of this chapter, PF project execution practices will be presented. The illustration below represents the Six-P model for managing PF projects.

Six-P Model for Managing Path-finding Projects

PF projects are defined from the vision and mission. While executing PF projects, however, the vision and mission of the organization are often reshaped. It is an iterative process of generating projects that align with the shared passion of the employees and the purpose of the organization. An organization develops its capabilities through learning, and PF projects are the means to obtain desired capabilities. It is a process often filled with uncertainties and failures. Understanding technology and market development cycles helps us to see through these deficits and executes the roadmap with patience and persistence.

Lastly, managing the execution of PF projects is different from managing Core projects and Continuous Improvement projects. Not all PF projects reach a successful conclusion but they all should create knowledge and teach lessons. Managing the performance of these projects and proliferating the learning from these projects are the key goals during execution.

Managing Path-finding Projects

Emerging Technology and Market

Most Path-finding (PF) projects will not produce successful products, but we should not consider them failures either. Without the path-finding effort, our world would not be advancing, especially in science and technology. Doing PF projects is part of process of exploration, which is the forefront of the journey toward innovation. This journey is never smooth and often requires patience and persistence. Knowledge of emerging technologies and market developments helps us to plan and execute PF projects.

In 1995, the IT research and advisory firm Gartner introduced the Hype Cycle for emerging technologies, and has been publishing a Hype Cycle chart each year since then. From 1995 to 2008, the Hype Cycle charts were plotted using visibility as the vertical axis and time as the horizontal axis. Starting in 2009, however, the vertical axis was renamed "expectation." The Gartner Hype Cycle shows the maturity of technologies in five phases: Innovation Trigger, Peak of Inflated Expectations, Trough of Disillusionment, Slope of Enlightenment and Plateau of Productivity [37].

The Hype Cycle should not, however, be used as a tool for predicting the trends of specific technologies. A study done in December 2016 revealed that among hundreds of technologies that have appeared on past Hype Cycles, only a handful of technologies followed the prediction of the Hype Cycle from start to finish [38]. Instead of focusing on predicting technologies, the Hype Cycle should be viewed as a general representation of technological advancement with respect to market expectations. The illustration below shows the Hype Cycle and its alignment of the technology adaption lifecycle.

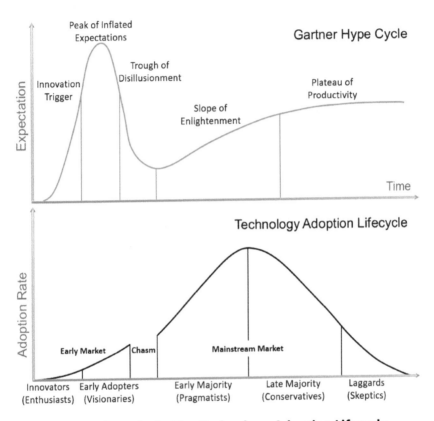

Gartner Hype Cycle Plus Technology Adoption Lifecycle

The technology adoption lifecycle describes the acceptance of a new technology over time for five demographic groups which were originated from a study of the Diffusion of Farm Practices by agricultural researchers [39]. The model has been adapted and further developed for many applications. The most popular application is a variation proposed by Geoffrey Moore in his book *Crossing the Chasm*

[40]. A disruption in technology adoption called the chasm is added to the previously continuous adoption curve.

The combination of Hype Cycle and technology adoption lifecycle explains that during the innovation trigger phase, the technology is typically adopted by innovators, who are the tech enthusiasts willing to try new things. As the technology gets publicity and moves into the peak of the inflated expectation phase, it gains the acceptance of early adopters who are visionaries trying to get ahead of others. The innovators and some of the early adopters create the early market for that technology.

As mentioned earlier, the study of 20 years of Hype Cycle reveals that most technologies simply disappear and die [38]. The technology adoption lifecycle with the chasm introduced by Moore explains this occurrence. The chasm typically occurs between acceptance by the early adopters and acceptance by the early majority group, aligning with the Trough of Disillusionment phase of the Hype Cycle. Visionaries tend to be more optimistic in seeing the benefits of a technology and downplay issues accompanied with implementation. This optimism serves to reinforce the value of the technology for this group, convincing themselves to adopt the technology earlier than others. The drawback is that this optimism typically pushes the technology to be hyped beyond reasonable expectations. Unlike tech enthusiasts and visionaries, the pragmatists only act when they see real value and generally do not trust visionaries. Moving into the mainstream market requires significant effort to cross the chasm.

Moore developed the concept of the chasm from the marketing viewpoint. In my opinion, marketing is only one of the factors for successful technology adoption. From the viewpoint of project management, PF projects are conducted in the early market phase and many of these projects fail not because they lack acceptance of tech enthusiasts and visionaries in the market. The illustration below reflects the modification of the technology adoption cycle from the project management viewpoint with project categories defined by this book. In the revised technology adoption lifecycle chart, disruptive technology is also added which is a technology that can take over an adopted technology in the mainstream market.

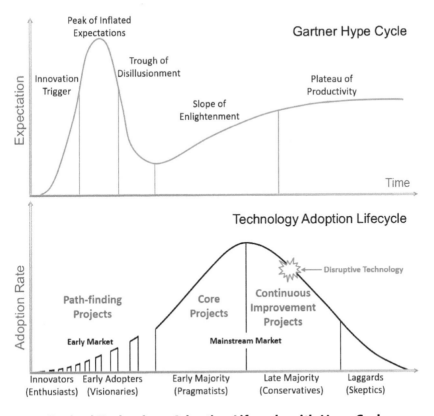

Revised Technology Adoption Lifecycle with Hype Cycle

Many of the technologies that disappear from the Hype Cycle did not just fall into one big chasm; rather, they disappear for a variety of reasons throughout the early market phase. These reasons include technical difficulties in development, changes in company strategy, inadequate funding and flawed approaches while managing these projects. There are, therefore, many pitfalls in the early market phase and not just a single chasm near the end. These pitfalls get bigger near the end of the phase due to declining expectations of the technology in the Hype Cycle as well as the increasing size of issues accompanied with market growth.

Taking a technology into the mainstream market is rough journey. It is a development process in gaining the necessary capabilities to overcome many pitfalls. Patience and persistence are required as the outcome depends on many variables and learnings from many areas of the business.

First, we need to learn the technology itself. Regularly reading the latest technical journals and publications in the given fields will provide insights into the technology and its trends. Reviewing the annual Hype Cycle published by Gartner also sharpens one's acumen in latest emerging technologies. Keeping track of current technology developments will increase our ability to recognize potential disruptive technologies that can significantly change the mainstream market.

Second, we need to learn about the early market and the behaviors of tech enthusiasts and visionaries. Innovation in PF projects means exceeding the expectations of these two technology savvy groups, not the expectation of the general market where most users are not technically oriented. Many PF projects fail to gain acceptance in the early markets because they focus on the mainstream market instead. Exceeding the expectations of the general market is the target for Core projects which will be discussed in the next chapter. To understand the marketing tactics required to reach the tech enthusiasts and visionaries, reading Geoffrey Moore's book *Crossing the Chasm* is highly recommended.

Lastly, we need to learn how to manage PF project execution effectively so that we can maximize the chance of success in the early market in order to reach the mainstream market, which is the discussion topic of the next section.

Managing Path-finding Projects

Creating Value and Knowledge

The exploratory nature of Path-finding (PF) projects often leads to unconventional approaches, unstructured processes and unexpected results. The performance of these projects are thus subject to a wide range of opinions. Some may consider a project a success while others may see it as a complete failure. The first key focus in PF project execution management is assessing the performance of these projects, which means understanding the values created by these projects. Conducting PF projects is also a learning venture, so proliferating the knowledge throughout the organization becomes the second key focus.

Performance Assessment

We begin the discussion with PF project performance assessment. To evaluate the success of PF projects, we must first understand what success looks like. Traditional Project Management uses the triple constraint to measure project success. The triple constraint is a project management model with three factors: performance, time and cost. Performance can be further separated into scope and quality. Under Traditional Project Management, a project is successful when it is done well per the defined specifications, on time and within budget. Using such measurements, most PF projects fail. We cannot therefore manage PF projects using the same success criteria based on the triple constraint concept.

Some people may argue that traditional project success criteria can still be used if the desired result or scope of the PF projects is tailored with learning objectives. In other words, a PF project is considered successful as long as the project achieved the learning objectives.

The first issue with this approach is that project success rate is inflated for the sake of measurement. The inflated project success rate influences risk-taking behavior by hiding failures; we want to encourage risk-taking by showing employees that it is okay to fail. Also, management needs to define the acceptance criteria for achieving the learning objectives, which is often difficult since learning occurs in almost everything we do. To what degree, however, is it worth calling the effort a success? That leads to vague criteria that hardly reflects real knowledge gain and evaluating these projects becomes merely an administrative formality.

The second issue is that setting project success criteria based on achieving learning objectives lowers the expectations of the team. Although conducting PF projects is for capability development, it is not a training program and should not be designed as one from the outset. The mental state of conducting actual projects is far more serious than that of participating in a training session. We want employees to pursue PF projects for real, aiming at actual technology and product development. If a project fails, it fails. Don't sugarcoat it by making it a consolation winner; that clouds the true achievements and makes real successes less exciting and meaningful.

Traditional Project Management sets time and budget targets for projects and uses these targets to evaluate effectiveness in managing these projects. Since PF projects are experimental in nature and have higher uncertainties, managing these projects with a set time and budget is not practical; we should not penalize the team if an extra week is needed for an experiment deemed very important or another thousand dollars for an extra prototype that significantly improves the previous one.

So how do we evaluate the success of PF projects? It depends on the purposes behind the particular project. PF projects can be further divided based on three main purposes: strengthening the vision, building capabilities and creating options for Core projects. In general, they will follow the timeline as shown in the figure below.

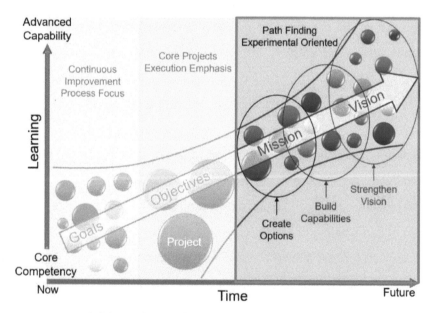

Division of Path-finding Projects by Purposes

Some projects may start to serve one purpose and then later transition to another, so it is possible for a given project to eventually serve more than one purpose. Managing tactics for these PF projects are sometimes quite different based on these purposes and a summary is shown in the following table.

| High Level Managing Tactics | Purposes of Conducting Path-finding Projects | | |
	Strengthen Vision	Build Capabilities	Create Options
Success Measures	Positive public image Funding attraction Talent recruitment Team enthusiasm	Patent Publication Industrial benchmark Employee initiative	Core project adoption Core team evaluation Executive grading Committee review
External Exposure	Publicity promotion Leading media coverage Visionary endorsement	Limited disclosure Conference participation	Confidentiality
Internal Visibility	Leader open forums Strategy update meetings Roadmap meetings	Idea submittal system Innovation events Employee development Educational seminars Demo and prototype Regular roadshows	Review meetings - Core project teams - Technical committee - Executives
Key Resources	Research labs Government agencies Industrial associations Universities Joint venture & partnership Outsource	Research labs Development centers Acquisition Partnership with NDA Employee free time	Research labs Development centers Acquisition Employee free time

Path-finding Projects Managing Tactics

The first purpose of conducting PF projects is to strengthen the vision of the company. Take the example of Elon Musk, the CEO of Tesla. Projects such as SpaceX, Hyperloop and Neuralink have significantly contributed to his distinguished reputation as one of the leading visionaries in the tech world. Conducting PF projects, especially the high-profile ones, authenticates his companies' visions and promotes their image. Being regarded as a visionary company helps the company attract investors as well as talented employees.

These projects often represent significant undertakings with huge challenges, so the chance of failure is extremely high. The success of these projects rests on their strategic purpose of conveying a vision to stakeholders and creating opportunities for the company. To test whether a project serves this purpose, we look at its ability to attract funding and hype from tech enthusiasts and visionaries as well as the team performing the work.

Managing these projects requires diligent attentions not only to internal activities such as technical development and resource management but also to external factors such as stakeholder relations and expectation management. These projects need to generate hype but cannot be overdone as they run the risk of being seen as hoaxes. Establishing credibility for these projects is very important and the selected project manager must have experience in managing external exposure to gain publicity, media coverage and endorsements from tech guru and visionaries.

Since these types of projects are very challenging and often require a long-term investment, leveraging external resources is highly recommended. Tactics include seeking funding and support from government agencies, collaborating with industrial associations and leading universities, working with other industrial leading companies though joint ventures or partnerships, and potentially even outsourcing to specialized firms. Project managers for these projects should obviously possess external relations and public affairs management skills that are less demanded for typical traditional project managers. For internal visibility, the practice of holding regular open forums and strategy update meetings to maintain the enthusiasm of the team and keep the vision vivid and inspiring is typically effective.

The second purpose of conducting PF projects is to build capabilities for the future. We should not expect a single project to bring in all the capabilities needed for the company; each project should focus on a subset of competencies, such as product design, materials development, processing technology, manufacturing methodology, etc. These projects are successful when the results are patentable and publishable, but patents and publications are not the only success measures. Achieving compatible results to industrial benchmarks with lower costs and fewer steps is also a success. Not all these results can be readily used on a particular product but collectively they enable the company to obtain a leadership role in its industry. Some projects may fail to achieve any desirable results but can still be considered a success if they spark employees' interest to initiate additional projects and delve deeper into an area.

More effort should be devoted to internal practices managing capability -building PF projects. Management should first provide a means to fuel employee initiatives. Being able to submit ideas freely through a submittal system and encouraging participation in innovation events such as internal hackathons and Shark Tank like programs are great ways to encourage this. Management should also align projects with employee development efforts and promote these projects as personal development opportunities in additional to presenting their benefits for the company. To maximize the success of these projects, related educational seminars should be provided regularly to employees. Finally, management needs to provide venues for employees to show their work and learn from each other. The external exposure of these projects, however, deserves cautious consideration. Due to patent applications and publications, limited disclosure of certain details to the public is necessary but key developments should be treated as protected intellectual property.

Resources for conducting these projects can be coordinated by the company's research labs or development centers which includes setting up special teams for specific focused areas. Acquisition is another way of obtaining a capability when internal skills are limited and time is critical. Partnerships with external companies are sometimes beneficial but non-disclosure agreements (NDA) should be in place to limit outside exposure. The above tactics are top-down approaches but some companies use the bottom-up approaches as well, such as allowing employees free time to start and work on projects as they wish. Post-it from 3M and Gmail from Google are a couple successful examples.

The third purpose of conducting PF projects is to create possibilities for Core projects. Core projects are essential to a company's success and possibly even survival. Jumping into a new area with a Core project, without the accumulation of capabilities gained from PF projects, is very risky. Conducting PF projects prepares the company to achieve success in Core projects with minimal risks. A good indication of success is thus the fact that some PF projects can transition to become new Core projects or adopted by current Core projects. Some PF projects may not be adopted immediately by an existing Core project but may have the potential for future adoption. Their success can be rated by Core project teams, the executive team or a review committee formed with the combination of technical experts and executives in the strategic office.

When managing a PF project that is coming close to becoming a Core project or adopted by a Core project, the details of these projects should be kept confidential to the external world and may even be limited to a portion of internal employees on a need-to-know basis. Exposing the details of these projects not only reduces the surprise element when the product is released to the market but also reveals the company strategy, enabling competitors to formulate responses ahead of time. Keeping the competition from guessing a company's internal strategy is, by the way, another reason for a company to conduct a large number of PF projects. Internally, the status of these projects needs to be communicated to Core project teams, technical committees and executives. The best way to keep these people informed is by setting up regular status review meetings. Resources used to conduct these PF projects are similar to the resources for projects building capabilities with the exception that partnerships are not preferred.

Success metrics and management tactics will obviously vary depending on the purpose of a particular PF project. As a result, the performance measures of PF projects are not as straightforward as those for Core projects. The success of a Core project is simply reaching the market with quality and velocity, which will be further discussed in the next chapter. Employees who work on Core projects are under a lot of pressure and failure is not an option. In contrast, employees who work on the PF projects are free to explore and performance is not measured by successful completions, permitting working in a more relaxed manner. This creates a situation where employees may feel it is unfair to be managed differently.

Managers of the PF project employees often face difficulty in evaluating the performance of the individuals working on PF projects. First, there are typically many PF projects in diverse fields making comparison of organizational impacts challenging. Second, these projects deal with the latest technologies in pioneering fields and the practical values are often not clear. Third, employees who work on these projects are experts and typically possess advanced degrees, making it difficult for managers to argue that the work of one individual is more or less of value compared to another employee. Therefore, when it comes to making decisions on promotions, pay raises and bonuses, it is difficult to do so fairly. If the individual performance evaluation system is not working, project performance evaluation will not be effective either. Project performance evaluation should align with the individual performance evaluations, so we should seek solutions by approaching both simultaneously.

We have discussed the cultural approach to management numerous times in earlier chapters. The essence of the cultural approach is making things visible and utilizing collective norms of employees to drive results instead of resorting to management control. The same principle can be applied in this case. Rather than management making an assessment, we utilize employees to evaluate the performance of PF projects and in turn translate that to individual performance assessments.

All PF projects teams must do a road show demo regularly. Quarterly or maybe shorter depending on the dynamic of the field and the industry. The demo projects are then evaluated using a standard evaluation form. If the confidentiality of the PF projects is not a concern, invitations to evaluate the projects can be open to all employees and potentially also outside experts such as university professors. If the PF projects are confidential, at a minimum, Core projects teams should be invited because PF projects may become future Core projects. Also, Core project teams are closer to the market and have a better understanding of customer expectations as well as the potential practical applications of the technology. Another less preferred option for evaluation is using a technical committee consisting of experts from Core project teams and PF project teams. Executives can also take part in assessing the strategic aspects of the PF projects, but it should not be the only consideration.

The design of the evaluation form should include the following areas: 1) the potential opportunity, 2) the innovativeness of proposed solution, 3) the adoptability and feasibility of the solution, and 4) the strategic importance. A scoring system should be used for rating these areas, and weighting may be applied to certain questions. Each PF project will receive a total score after the assessment. For individuals on the PF project team, a peer evaluation is done with each team member to assess the contributions of others. The combination of the PF project evaluation score and individual peer evaluation score determines the employee performance, which serves as the basis for promotions, salary increases and bonus decisions.

Management should not take hasty actions when a PF project receives lower scores occasionally, especially in the beginning stages of the project. Consistently lower scores over several periods, however, may warrant a discussion with the team. Typically, a PF project with a solution that is readily adopted by the current Core project will score higher. A PF project will score lower when the solution and its practical applications are difficult to understand by the evaluators. Sometimes it is because the technology involved in the PF project is so advanced that only a few can comprehend. In this situation, we simply direct the PF project team to go work with the evaluators, educating them on why the project deserves higher scores. This topic then leads to our discussion of the second key focus on doing PF projects: knowledge proliferation.

Proliferation of Knowledge

Many companies have difficulty capturing business value from their research and development efforts. In essence, they have difficulty transitioning from PF projects to Core projects. It is a common practice in many companies to conduct R&D through dedicated organizations, such as Bell Lab, Xerox PARC, GE Global Research, HP Labs, Intel Labs and Google X. Utilizing a dedicated research organization to drive R&D certainly demonstrates that the company is very serious about the path-finding effort. The common problem, however, is communication between the research organization and the rest of the company. In most companies, the research lab is operating within its own silo and the main communication link is through management. The managing director of the lab communicates upward to the executive office then the executive office communicates to the heads of the other functional organizations. This may be fine for strategic coordination and decisions, but the technical capabilities developed from PF projects remain stuck in the research organization and, as a result, technical knowledge is not proliferated to the rest of the company.

It is also common for companies to use a PF project team to continue to manage the project when it becomes a Core project. Typically, a new organization or team is formed outside of the research organization, and its expansion and operations are led by the personnel from the PF project team. From a technical perspective, it makes sense as these people developed the solution. However, it is not ideal from a managerial perspective.

First, the knowledge continues to be held by only a few key individuals and dependence of these key individuals may cause issues for the company. Second, managing Core projects requires a significant change in mindset from conducting PF projects. PF projects require outside-the-box thinking but Core projects demand discipline and methodological execution. Although a bit of an exaggeration, this is analogous to putting a university professor in charge of a company's operation. I apologize to university professors who may feel offended, but most research people cannot, in a short amount of time, shift their mindset and gain the necessary skills to manage operations successfully. Core projects are very important to the company and must reach the market successfully. It is too risky to treat them as extensions of PF projects. The details of managing Core projects will be discussed in the next chapter.

The preferred approach to transitioning PF projects to Core projects is to transfer the knowledge instead of people. Management must create venues for engineers on PF projects to work with engineers on Core projects. That is why regular road show demos are required and the participation of Core project teams is necessary. These not only serve as performance evaluation mechanisms for the PF projects but also learning opportunities for the Core project teams. If done correctly, the knowledge sharing does not just happen in the demo sessions; management should encourage members from PF projects that receive lower scores to contest those results to the Core project teams, which typically leads to the PF project teams educating the Core project teams on the theoretical aspects of the development while the Core project teams explain the practical aspects to the PF project teams. By exchanging viewpoints, the PF project teams will most likely adjust their focus while the Core project teams will have a better understanding of the future trends and be able to prepare for the next challenges. The result is a smoother transition from R&D efforts to Core project execution and, more importantly, organizational level of learning is achieved through knowledge proliferation.

Planning and Executing PF Projects

With clear mechanisms for performance evaluation and knowledge proliferation, we shift our attention to the logistics of managing PF projects, specifically the planning and execution practices. In Chapter 3, when MBPM was introduced, we mentioned that Extreme Project Management, which was introduced in Chapter 2 Page L2-6, is more suitable for managing PF projects. It is rarely practiced in companies, however, as current models are ambiguous. I have previously shared my view that we should not attempt to create a comprehensive model with concrete processes for Extreme Project Management as we want to remain flexible and allow creativity and outside-the-box thinking. However, this does not mean that these projects are done in a relaxed manner and managed without much discipline.

An analogy used to explain Extreme Project management is pursuing a doctorate degree. Getting a doctorate degree is characteristically challenging and about 50 percent of doctoral students quit the

programs without getting the degree [41]. It is not just about coming
up with new ideas but the process of developing and proving the ideas.
Doctoral students are under tremendous pressure, and from my
experience in obtaining multiple doctoral degrees, those who are
disciplined in the process have a higher chance of success. In 2015, I
was asked by UC Berkeley to design a course to help Ph.D. students
increase their chances of getting the degree. I found that the approach
exemplifies Extreme Project Management and is applicable to managing
PF projects in the business world.

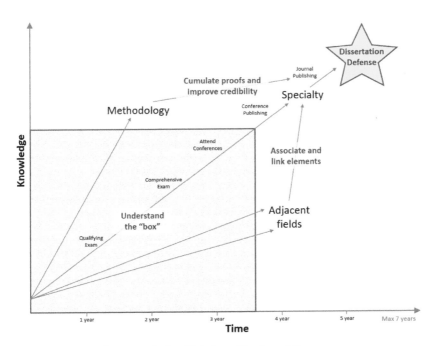

Approach to Obtain a Doctoral Degree

The above illustration shows the approach that I recommend for
doctoral students which is a process of accumulating knowledge over
time. The entire program is time-bound with many milestones that must
be met under relatively strict timelines. Most universities have a 7-year

limit to finish the program. The majority of graduate students find doctoral programs quite intense with many activities to be done in a short period of time. The illustration below shows recommended activities for the first year.

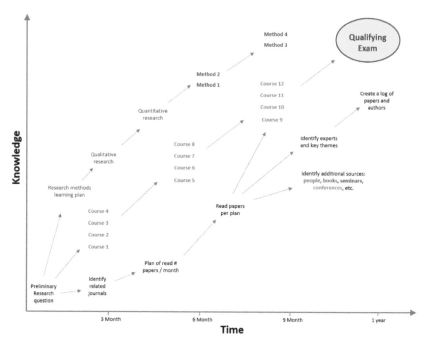

First Year Activity Plan for a Doctoral Degree

In addition to the coursework needed to reach the first major milestone, the qualifying exam, many other research activities are necessary to ensure the success of the entire program. Many doctoral students made the mistake of going through the first couple of years of doctoral programs as bachelor and master programs by only focusing on coursework. They would typically face difficulty in meeting the research and publication timelines in the later part of the program. Coursework and research are interrelated and the knowledge gained in courses supports the research.

Going through a doctoral program is clearly intensive and demands early planning on key activities. Likewise, when doing PF projects, we need to plan ahead with an overall time target and frequent milestones. Through regular reviews and grading, similar intense pressure may be applied to the PF project teams.

The first part of the process is to understand the "box." The "box" refers to the existing knowledge in the field. Not only does the subject specialty need to be learned, the methodology used to prove the current theory must be studied as well. In fact, the learning curve of the methodology needs to happen at a faster pace. Based on the Triple-loop Learning Theory discussed in Chapter 4 Page L4-22, effective learning occurs when the underlying assumptions and the context of the theory are fully understood. Knowing how others derived the theory helps students to comprehend the theory as well as develop strategies to prove their own theories.

In addition, learning theories in the adjacent fields will broaden their perspectives; many innovative solutions are a fusion of technologies from multiple fields. Similarly, when a PF project is initiated, the first phase is to learn the existing solution "box," which includes all the existing practices, the methods used to develop these solutions, and practices in similar fields. For instance, if we are developing a better cleaning method for semiconductor manufacturing, we should research all the existing cleaning methods, how each cleaning method is developed and processed, and the cleaning methods from other industries that also require cleaning, such as cleaning methods used in surgery, dentistry and maybe even archaeology.

The process of understanding the 'box" consists of rigorous research and requires disciplined planning and organization. It is a process of building a database of existing solutions. Until the "box" is fully comprehended, no one can be certain that a new idea is truly outside-the-box. Sometimes, we will find a proposed solution has already been done by others, or an existing method actually produces better results than the proposed solution, all of which are normal. We will then seek the next solution or drop the PF project. Only the projects that have potential to go outside of the "box" will continue to the next phase. Therefore, it should be a formal phase exit similar to the comprehensive exam in the doctoral program.

When PF project teams believe that they have a full understanding of the "box" and are ready to present the findings, they can request a review session. A review committee is formed with key Core project team members such as chief engineers and product owners, technical experts such as fellows and principle engineers, as well as executive members of the company, serving a gate keeping function similar to that of a doctoral committee. If confidentiality is less of a concern or NDAs are in place, outside experts such as university professors can be invited as part of the committee to ensure the work is in fact outside-the-box.

Once a PF project passes the gatekeepers and exits the "box," the review committee will work with the PF team to set the ideal end target date of the project as well as the timeline for achieving that target. Regular review sessions continue to occur, setting and tracking to the milestones such as publication, patent application and Core project adoption. In the academic world, another function of the doctoral committee is to provide guidance and resources for the student. Similarly, regular review sessions serve as an escalation channel for the PF team to seek help. These regular review sessions are not the same as the evaluation sessions. The evaluation sessions are for performance assessments similar to grading in school every term. Regular grading of PF projects ensures the performance of the project and team members is sustained.

The evaluation sessions for PF projects involve a bigger population, such as all Core projects teams in addition to the review committee, and occur less frequently. The recommendation is to hold evaluation sessions quarterly and review sessions bi-weekly or monthly. These regular interactions provide a venue for PF project teams to demonstrate their work. If a PF project is truly interesting with big potential, it will shine and should not have problem in attracting resources, just like how novel entrepreneurial ideas attract investment. By design, Core project teams are bigger and have more resources. Attracting Core project teams to engage in the PF projects helps the company to turn its R&D efforts into real business value.

A checklist is provided to help capturing all the main actions for managing PF projects. In summary, PF projects are managed through the Six-P model:
- Passion drives the motive in pursing PF projects.
- Purpose provides the strategic direction for the path-finding effort.
- Patience is needed as technology development takes time and we are all students in this pursuit.
- Persistence is also needed as there are many pitfalls early in the adoption cycle.

- Performance evaluations need to be done regularly and fairly by utilizing the cultural approach to management.

- Proliferation of knowledge must occur to ensure capabilities are transferred to the Core project teams.

Checklist for Managing PF Projects
☐ Develop a vision (from passion)
☐ Define a mission (purpose statement)
☐ Identify targeted areas for learning
☐ Understand key technologies and their current lifecycle stages
☐ Initiate and map Path-finding projects based on purposes
☐ Strengthen vision
☐ Build capabilities
☐ Create options
☐ Develop high-level managing tactics
☐ Success measures
☐ External exposure
☐ Internal visibility
☐ Key resources
☐ Setup review committees
☐ Core project teams
☐ Key members, e.g. technical experts, executives
☐ Schedule review and grading cadence
☐ Conduct project review and grading sessions
☐ Performance grading
☐ Core project adoption potential
☐ Patent application and publication possibility
☐ Learning and sharing forums knowledge proliferation

Checklist for Managing Path-finding Projects

Managing Path-finding Projects

Exercise Questions

E-mail your thoughts on the questions to MBPM.Innovation@gmail.com
I will share my thoughts and answers to the questions.

Rules: (For details and reasons, please read Preface Page xvii)

1) One question at a time and state the Question # in the subject line
   of the email.

2) Provide a scanned copy of the book purchase receipt the first time
   you use an email to send in a question. This won't be necessary for
   future questions using the same email.

3) My response will be sent to you between 2-4 weeks after I receive
   your email.

Q8-1. A company's vision and mission guide the path-finding effort.
       When do you think a company should review and update its
       vision and mission? What are the subsequent actions for
       realigning path-finding efforts?

Q8-2. Following the latest news in technology, the technology
       development cycle seems to be shorter and shorter in most
       industries as new technologies appear more frequently. On the
       other hand, technological development becomes more and
       more difficult and requires patience and persistence. Are we in a
       dilemma and how should we cope with this?

Q8-3. PF projects are conducted in the early phase of the technology adoption cycle. Success in the early market does not guarantee success in the mainstream market as the customers are different. From the customer and market perspective, what can a company do to increase the successful transfer to the mainstream market?

Q8-4. PF projects for strengthening vision are often strategic and initiated from the top. Is evaluation by Core project teams necessary? Should the evaluation be done by the executives and the chief engineers only?

Q8-5. It is not recommended to transition a PF project team into the Core project team (Page L8-11 to L8-12). How should we form and organize the team for the PF to Core project transition?

Q8-6. From what you read in this chapter, what criteria would you use to select PF project managers? Are the skill requirements the same for all project managers in general?

Chapter 9

Managing Core Projects

Prioritization and Focus

Core projects aim for the next big market entry which will be the next generation of products or services offered by the company. They are the most important projects among all projects and they need to be successful in order for the company to be successful. As shown in the MBPM strategic chart below, there should only be a few Core projects so that the company's resources are used effectively and efficiently. Prioritizing and having a clear focus are therefore extremely important when managing Core projects.

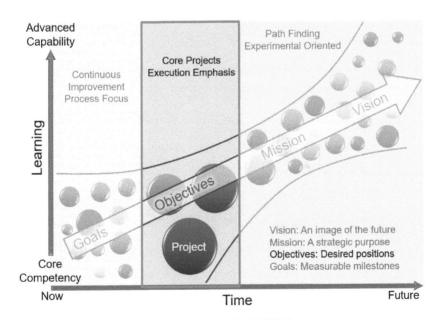

Core Projects in MBPM

In the T section of last chapter, we discussed the initiation of Path-finding (PF) projects, which are derived from a company's vision and mission. Likewise, the initiation of Core projects is discussed in this section, which is more of a selection process based on the company's objectives. There are two general objectives as shown in the illustration below, which drive different Core project selection tactics: growth and survival.

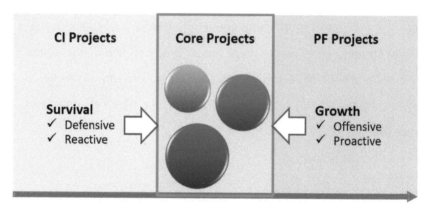

Defining Core Projects Based on Survival or Growth

If a company's objective is survival due to a degrading financial or market position, management should define Core projects by assessing existing Continuous Improvement (CI) projects. That means expanding or combining CI projects to become Core projects, such as making significant improvements on an existing product to become the next new product. Taking a defensive position is not always the most desirable course of action, but often we need to defend until we have the strength to attack.

On the other hand, if a company has already achieved financial and market stability and its objective is growth, its Core projects should be selected from PF projects as described in the last chapter. This proactive approach is preferable as it takes the offensive position. Even if a company is being pushed into the defensive position, it should still be looking into the possibility of transforming a PF project into a Core project. An organization should ultimately target growth instead of survival. Referring to the career development analogy that has been used several times previously, an individual must aim at career growth rather than just surviving in the current job, although in the short term sometimes keeping the current job can be the highest priority.

As a result, a company may execute several Core projects, some targeting survival and some aiming at growth. A focused strategy must be carefully derived as typical companies have limited resources and cannot afford to simultaneously execute too many large projects. We mentioned earlier that Core projects take priority over all other projects, including PF and CI projects. Among the Core projects, they should be clearly prioritized as well so that everyone in the organization knows which project is Job#1, Job#2 and Job#3. It is like an express train such that when it arrives all other trains must yield. Large corporations have many divisions so each division can have its own Job#1. Nevertheless, I still believe that an organization as a whole should have a master priority list; if not, the divisions may operate as silos and coordinated effort towards the organization's objectives will be lacking.

Whether the Core projects are derived from PF or CI projects, the selection process must be objective. These projects are important to the company and because Core project managers are in charge of a large portion of the company's resources, office politics is often present. Avoid autocratic decisions as they will most likely reduce team commitment, which is extremely critical for Core project execution. A team decision is preferred and tools such as Brainwriting, Brainswarming and Six-step Decision-making model described in Chapter 5 may be used.

In the M section of last chapter, we discussed that patience and persistence are required when conducting PF projects. In the M section of this chapter, which is the upcoming section, we will discuss the key emphases for conducting Core projects, which are quality and velocity. In the L section of this chapter, we will discuss the key practices for Core project execution.

Chapter 9

Managing Core Projects

Quality and Velocity

Core projects produce real products and services, so quality is essential. Poor quality will be much more costly to fix when these products or services reach the market. These projects not only need to attain successful results but also must be completed quickly. Quality and velocity are the key objectives when executing Core projects.

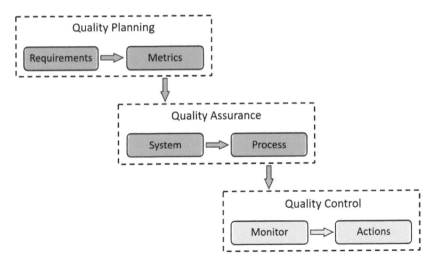

Traditional Project Quality Management

As shown above in Traditional Project Management, project quality is managed through three main processes: quality planning, quality assurance and quality control. Quality planning is identifying the quality requirements and establishing metrics to measure them. Quality assurance is ensuring the accuracy and credibility of the measurement system and processes. Quality control is executing the processes by monitoring the metrics, identifying the gaps and taking preventive or corrective actions.

There is nothing wrong with the traditional project quality management approach but it generally requires a lot of administrative effort. For Core projects, we want velocity so we must keep things simple and reduce non-value added work. Under Total Quality Management (TQM) principles, quality is a culture and everyone should pay attention to quality when they are executing tasks. Applying TQM, quality assurance and quality inspections are wastes, as the ideal state is to build quality in every step so that quality checks are unnecessary. Establishing a culture of quality is critical when managing Core projects.

A culture of quality starts with leadership and an emphasis on the importance of quality from the top. Management must clearly communicate to all employees the significance of producing quality work. However, simply making a request is not enough. Many companies have done this and the results have been less than desirable. It is very difficult for a person to do everything perfectly since humans have attention span limits. Clearly prioritized tasks allows employees to know when to pay special attention to ensure these tasks are done correctly the first time. For the Core project team members, the WBuS task management approach described in Chapter 6 and the Critical Chain scheduling method described in Chapter 7 provide clear prioritization and focus. As mentioned before, in general lower-level tasks in WBuS have higher priority than the higher-level tasks. Tasks that are on the critical path also receive higher priority. It is recommended to assign a quality requirement rating to each task based on WBuS and Critical Chain Scheduling. The details of how to do that will be presented in the next section. With a clear quality requirement rating system, employees will know when to be careful and when they may relax. For support personnel, Core projects take the highest priority over all other projects and the focus on quality must be higher.

Emphasizing the importance of quality from the top and communicating requirements are yet still not enough. A mechanism in quality performance management must be in place to ensure the execution of those requirements. In order to be effective, such a mechanism must be designed based on the principles of the cultural approach to management. A simplified quality management approach is shown below.

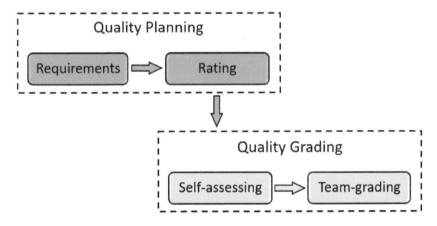

Simplified Quality Management for Core Projects

Immediately after the completion of each task, a quality grade is assigned to the task. The process should be self-graded first by the task owner and then reviewed by a few team members before assigning a final grade. The ideal reviewers are the owners of the subsequent tasks as they are the ones who are impacted the most. Managers should not be involved. Quality requirement ratings and grades should be displayed openly to drive performance through peer pressures.

Velocity is another key focus when managing Core projects. Management must create a sense of urgency for Core projects, not just within the Core projects teams but across the entire company. At the company level, executives must communicate often so that everyone in the company knows the Core projects and their importance to survival and growth. Special attention and treatment is given to the Core projects. The Core projects teams and members receive undivided support when needed. They are like an arriving express train; all others trains must yield. Management needs to set an example for this commitment. When it comes to decision-making related to the Core projects, they must demonstrate decisiveness and quick action to seek results. The RAPID decision-making and CREATOR meeting management models described in Chapter 5 can be utilized to speed up the process.

At the team level, Core project teams must be operating with a fast tempo by utilizing the Critical Chain Method for time management as well as Agile Project Management. Tasks, schedules and performance must be visible to drive accountability. In addition, there should be incentives for completing tasks early. Frequent and regular status review sessions are needed to drive momentum to ensure actions occur at fast pace. Core project team members are under pressure to perform so psychologically they need to feel important and motivated. Core team leaders and members must be selected carefully. Being a member of a Core project team should be considered an honor. Appropriate measures and practices should be implemented to make Core team members feel special and to enable them to work at a fast pace without typical administrative routines and burdens. The next section will present details for building a Core project team and implementing measures and practices to enable these teams to operate with the highest possible quality and velocity.

Managing Core Projects

Executing to Achieve Results

Core projects are the most important projects of a company and they take priority over all other projects. There should also be priorities within the Core projects so that all employees know which one is the most important when multiple Core projects require simultaneous attention. As mentioned before, there should be only a limited number of Core projects so that the company can effectively utilize its resources to execute a focused strategy. For large companies with multiple divisions and market segments, priorities are set at the division level. In my opinion, there should be no more than three Core projects being executed at any given time. Clear priorities are communicated to the entire organization in the form of Job#1, Job#2 and Job#3.

Forming Core Project Teams

Once the Core projects are selected and prioritized, the next task is to build the Core project teams. These teams are special and operate under exceptional pressure to perform with high quality and velocity. Depending on the characteristics of the Core projects and the corporate culture of the company, three options may be considered in building a Core project team:

1. Build the Core project team like a world-class sports team.
2. Build the Core project team as a Special Force or SWAT (Special Weapons And Tactics) team used by military and law enforcement agencies.

3. Build the Core project team as a scrum team by utilizing the Agile Project Management approach, which was introduced in Chapter 2.

Regardless of which option is used, Core project team members must be selected carefully. When selecting the Core team members, I would prefer to think along the lines of building a sports team or a SWAT team by paying attention to personality of the candidates. Not all people are suitable for such a fast-paced and high-stress work environment. These teams need members who possess not only excellent technical skills but also personality traits such as conscientiousness and cooperativeness.

Based on a study conducted on 34 SWAT teams by Deanna Marie Putney of the University of Tennessee, individual conscientiousness and agreeableness are significantly and positively correlated to team performance and viability [42]. Putney observed that individual conscientiousness consists of six facets: order, competence, achievement striving, dutifulness, deliberation and self-discipline. Agreeableness, meanwhile, has six facets: straightforwardness, trust, tender-mindedness, compliance, modesty and altruism. While editing this book, Mark Cramer pointed out that honesty should be included as well. I believe that honesty is the foundation for trust. Forming Core project teams with individuals who possess these personality traits helps to create an action-oriented norm that enables high-quality and high-velocity execution.

Selecting the right leaders for Core project teams is even more critical. They need to have not only the right personality and management skills but also reputation and respect within the company. We could call these leaders project managers but I prefer building Core project team as a sports team. Each Core project team needs a coach and a captain. Both are leaders, but they serve different functions.

A coach focuses on setting a winning strategy, assessing performance, developing team skills, motivating the team, mentoring team members and cheering for their efforts. The Core project team coach is typically management staff.

Meanwhile, a captain is typically a respected technical leader of the Core project team. He or she can be a star player, but this individual must put the team first instead of his or her own performance. The individual must also be a role model with conscientiousness and cooperativeness plus responsibility and decision-making in his or her area of expertise.

This dual-leadership model has obvious advantages since it clearly defines the two roles in covering the necessary responsibilities of a team, avoiding the dilemma in typical project management on whether to select the project manager from either the management staff or the technical ranks. When the decision is made to go with a single project manager, some of the responsibilities may lack ownership or receive less attention. A project manager selected from management staff may not have the technical knowledge to make fast and good quality decisions. On the other hand, a project manager from technical ranks has the tendency to be task-oriented and may pay less attention to the functioning of the team. The similar model is used in the Agile Scrum approach where each Scrum team has a Scrum Master and a Product Owner. The Scrum Master serves the role of a facilitator while the Product Owner is responsible for the clarity of the work.

As mentioned in Chapter 3 when MBPM was introduced, Agile Project Management is suitable for managing Core projects. More specifically, Core projects are best managed under the Agile framework, but current Agile approaches need to be modified to enhance Core project success. Most of the current mainstream Agile methods have excess administrative functions and rules: the Scrum Master and Product Owner, for example, both are not doing the actual work. They are overhead by definition. In addition, the success of Scrum requires support from a manager, who is not part of the team. In my opinion, there are too many chiefs and excess administrative overhead.

I agree that we need leaders in both managerial and technical areas. A single project manager generally does not cover both areas adequately. That is why I prefer building Core project team like a sports team with a coach and a captain. Comparing to Scrum, the captain takes on technical leadership as the Product Owner but also does actual work and is a role model while executing tasks with quality and velocity. The coach takes on a combination of the Scrum Master and manager responsibilities, as well as the Product Owner role by defining the "What." Scrum also has too many rules, many of which are too rigid. Keeping the process simple and reducing administrative overhead will certainly increase the quality and velocity of project execution. We will discuss further the process simplification later in this chapter.

Developing Core Project Teams

Once team leaders and members are identified, the team needs to complete the forming phase and quickly move to the performing phase. According to Tuckman's team development model, typical teams go through four stages: forming, storming, norming and performing [43]. During the forming stage, team members display wide range of behaviors depended on their personal interests, objectives, expectations and personalities. Some members may show eagerness and excitement, while others are confused and hesitant. Team interactions are generally superficial and polite. As the team moves into the storming stage, conflicts emerge as a result of the team's effort to institute rules and orders. Individuals seek power and influence to enact rules in their favor and take on preferable roles and responsibilities. Once the conflicts are resolved and the dust settles, the team moves into the norming stage. Acceptable behaviors and team rules have been established and team members consent to their positions in the team. The team exhibits cohesion and harmony in attitudes, values and expectations. At this point the team is finally ready to move into the performing stage where members collaborate and delegate freely as well as demonstrate accountability with minimal supervision.

According to Tuckman, each stage builds on the previous and skipping any stage affects performance negatively. Due to the focus on velocity when executing Core projects, there is a tendency to prematurely put the team into execution mode. This creates issues and conflicts as the team will loop between the stages and take even longer to perform effectively. Understanding Tuckman's model helps us to plan actions to guide the team through all the stages to reach the performing stage as quickly as possible.

Letting the team go through the forming stage by itself could take some time, however. Leaders of Core project teams should take an active and directive role to move the team out of this stage. The intention is to encourage all team members to open up and share their interests and expectations. Icebreaker and team building activities such as MBTI [44] and DiSC [45] are often used. These activities, however, are typically non-work related and the time spent would not be effective. In addition, these activities often give employees an impression that management does not want to take time to understand them, and instead, lets them fill out a survey to share who they are. People often put their guards up and fill out the survey with what they think management would like to see or otherwise based on a non-existent version of their ideal selves. After doing these exercises time after time whenever a project team is formed, the results often vary and are less meaningful in revealing their true personalities.

My recommendation is to utilize project logistic tasks as team building exercises, which serves a dual purpose. Project War Room setup, for example, can be a good team building exercise. The details of setting up a Project War Room will be discussed later in this section. While doing these tasks, the team leaders should consciously help the team build rapport by injecting questions and comments relating to personal characteristics and interests: "How do we feel about putting this table here?" "Why do you like this setup and tell us more about it?" "Now we know what you like." "We need the detail-oriented people to design this board and the creative artists to do that poster." Casually, the personalities, interests, strengths, skills and backgrounds of the team members are revealed, achieving the same purpose as non-work-related team building exercises but in a setting that allows the team members to reveal their true selves. Making these tasks fun with casual jokes and humors will also increase the effectiveness in building trust in the forming stage.

Once all the team members are comfortable in discussing issues openly and are committed to the purpose as a team, team leaders need to move the team into the storming stage. Start this stage by discussing individuals' roles and responsibilities and seeking commitment to the tasks and processes. Developing the WBuS with task assignments is the right exercise for this stage.

Not all teams go through a tense storming phase. For Core project teams, if we select members with the right personality traits as discussed earlier, the storming phase will be mild. Most individuals are more likely to compromise if they believe what they are doing is important and for a great cause which, by definition, all Core projects should be. Since Core project teams have a reputation of being special and it is an honor to be part of these teams, members tend to be more willing to make personal sacrifices.

When conflicts surface, team leaders need to ensure issues are confronted openly and professionally by encouraging members to express different opinions and reinforcing good team behavior. There is a useful rule of engagement in managing team conflict presented in the book *The Advantage* by Patrick Lencioni: silence is interpreted as disagreement by the team leader [46]. Practicing this rule, I found it quite effective especially when it expands beyond the leader with team members calling each other out when they are silent, pushing each other to express and commit to the discussion.

When the team resolves differences and all members are clear on their roles and responsibilities, the team moves into the norming stage. Building consensus on processing and planning issues is the right target for the team at this phase with appropriate activities such as developing task schedules and defining quality ratings. I recommend using the Critical Chain Scheduling described in Chapter 7 as the task schedule is done by the team rather than by the project manager. Another alternative is to use the Scrum model with the team working together to build the epics, features and stories. Both approaches require active participation from all team members, which serve as a good medium for strengthening the team during the norming stage.

Task	1.1 5.1 B.1	2.1	1.2 C.2	2.2 6.1	2.3 5.2	1.3a A.1a	1.3b A.1b	6.2	2.4	6.3	6.4 A.3	2.5 5.3 B.2	2.6	6.5 5.4	2.7 6.6	5.5
Duration	7	4	5	5	6	4	5	4	5	5	5	4	4	4	5	4
Ideal Start	0	0	3.5	3.5	6	6	6	6	9	9	11.5	11.5	13.5	14	16	16
Target Start	0	0	6.2	6.2	10.7	10.7	10.7	10.7	16.1	16.1	20.5	20.5	24.1	25.0	28.5	28.5
Actual Start																
Actual End																
Quality Rating	A	C	A	A	A	B	B	C	A	A	A	B	C	A	A	C
Quality Grade																

Owner 1   **C.3** 2d D   **B.3/A.4** 3d D

C.1   1.2/C.2   2.3/5.2   6.3   **2.5/5.3/B.2**
3d D   2.5d A   3d A   2.5d A   2d B

Owner 2   **C.4** 2d D

1.1/5.1/B.1   2.2/6.1   **1.3a/A.1a**   **2.6**   **5.5**
3.5d A   2.5d A   2d B   2d C   2d C

Owner 3   **A.2** 2.5d D

2.1   **1.3b/A.1b**   **C.5**   6.5/5. 2.7/6.6
2d C   2.5d B   2d D   2d A   2.5d A

Owner 4

6.2   **2.4**   **6.4/A.3 C.6**   **A.5**
2d C   2.5d A   2.5d A   3d D   2d D   14.5d

18.5   33

Legend: **Task**   Critical Path   Buffer   **Free Task**   Duration Quality Rating

Critical Chain Schedule with Task Quality Ratings

As discussed in the previous section, quality is one of the two important focuses in managing Core projects. Therefore, quality ratings should be assigned to each individual task. If Critical Chain Scheduling is used, quality ratings can be added to the schedule and table views as shown.

Continuing with the example from Chapter 7, quality ratings are added to the previous Critical Chain Schedule Table (Page L7-15) and the Schedule View (Page L7-16).

The team should work together to assign and agree on task quality ratings. Tasks in the critical path generally have the highest "A" rating because any rework would affect the entire project schedule. The table also includes a row for quality grading which will be done when each task is completed. The approach to grading was discussed in the previous section on Page M9-3. The task owner should start with self-grading followed by peer grading by task receivers and technical leads. If there are discrepancies in the grades, the task owner needs to discuss the reason with the peer graders to agree on a final grade. If they cannot come to an agreement, it should be brought up to the team for open discussion and judgment. Quality grading is obviously done at the performing stage, but the grading process is defined and agreed in the norming stage.

Utilizing project logistic and planning activities instead of the typical team building exercises takes the team through the forming, storming and norming stages to the performing stage quickly, creating a sense of urgency that accelerates the velocity in executing Core projects.

Moreover, this sense of urgency towards Core projects must not only exist with the Core project teams but also needs to be extended to the entire company, which reflects the priority and special treatments given to the Core project teams and their members. For example, dedicated parking spaces close to the building are given to the Core team members. There is a special line for them in the café, or even better, food is delivered to them. They have their own conference rooms and private rooms. Each Core team member is issued a special badge that grants them access to these privileges, which increases the visibility of the Core projects so everyone in the company knows and gives priority when requests come from these teams are sent to purchasing, logistics, legal, HR or to management.

These practices accelerate Core project velocity by aiming at eliminating any wasted time or inconvenience for the Core project teams and their members. It is special to be a Core project team member and we want to attract people to be one. However, an individual must have the qualifications and be willing to work hard to deserve these special treatments and the honor of carrying that badge. Viewing this from a different angle, these practices reduce excuses from the Core team members for not performing with velocity. Letting the Core teams generate their wish lists can be an exercise in the team building process. Wishes are granted based on the company's culture and availability of resources.

Dedicated Project War Room

Another privilege is a dedicated Project War Room which must be given to each Core project team and preferably only to the Core project teams. It makes it convenient for the team to plan, work and meet at the same location without wasting time on logistics. The dedicated War Room not only gives the Core project team an identity and a sense of belonging but also creates a connecting environment by surrounding the team with information regarding the Core project. It is a special room requiring a careful design. Below is an example of a Project War Room designed for a small team.

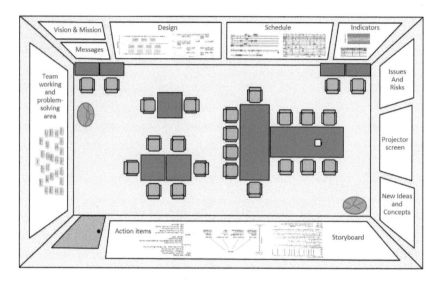

Project War Room Design Example

Designed correctly, a Project War Room can handle a lot of information, allowing cross-referencing and simultaneous problem-solving. Project management administrative tasks are reduced. Information is easily rearranged so plan revisions using computerized tools can be reduced or even eliminated. Since project information is updated real time along with the progress, management should visit the War Room for project status updates instead of requiring the team to put together PowerPoint presentations to show during a meeting. In other words, management should also use the War Room as much as possible to avoid burdening the team with administrative tasks so that they may focus on their work. For instance, if a customer requests a status update, management should go to the War Room, take pictures of the needed information and present them in PowerPoint to the customer without bothering the team.

War Room Design Component Checklist
❑ Vision and mission
❑ Key objectives
❑ Members and roles
❑ Announcements and key communications
❑ Message board
❑ Reminders
❑ Rules of the War Room
❑ Project plan and status update
❑ Design diagrams
❑ Process maps
❑ Storyboards
❑ Schedule
❑ Budget
❑ Issues and risks
❑ Action items and to do list
❑ Problem-solving Area
❑ New ideas/concepts space

Project War Room Design Component Checklist

There is no need to hire a design firm or a consulting company to design a Project War Room. Shown above is a War Room design component checklist. The War Room is for the team so it is best to let the team determine its setup, which should be a team building exercise as mentioned earlier. Some general guidelines can be given to the team but ultimately, the team owns the working environment as its members will spend countless hours in the room. A checklist for a War Room facility setup is shown on the next page.

The company facilities support team should ensure the War Room is set up and decorated based on the team design. Security badge access is installed so only the team and a few support staffs have access to enter. Provide plenty of snacks, healthy alternatives and beverages and restock them regularly. Deliver food to the room for the team and clean up after each meal. Mail and packages are delivered to the room as well. In short, the Core project team deserves priority support so that the team can focus on completing their work with quality and velocity.

War Room Physical Facility Setup Checklist
❑ Large main table with chairs
❑ Small semi-private work spaces
❑ Small collaboration or huddle spaces
❑ High-tech tools
❑ Computers
❑ Monitors and displays for laptops
❑ Projector and screen
❑ Teleconferencing system
❑ Network/wireless connectivity
❑ Low-tech tools
❑ Whiteboards
❑ Markers with different colors and easel pads
❑ Different size sticky notes, color stickers and sticky flags
❑ Office supplies: notebooks, pencils/pens, etc.
❑ Wall space usage
❑ Cork bulletin boards, magnet boards, or foam boards
❑ Mounted white boards, or erasable wall paint
❑ Tape, tacks, pins, removable adhesive, hanging strips, etc.
❑ Comfort and convenience items
❑ Kitchen appliances: coffee machine, microwave, refrigerator
❑ Snack cabinet, candy baskets, cups, utensils, etc.
❑ Coffee, tea, snacks, candies, drinks, etc.
❑ Toys and games for creativity and stress relief
❑ Couches, sofa beds, beanbags, etc.

Project War Room Physical Setup Checklist

To use the War Room effectively, a set of operating rules needs to be established. Again, these rules should be generated by the team as a team building exercise. Any disagreement over the rules needs to be openly discussed during the storming stage. The final version of the rules needs to be accepted by all team members and then enforced by the team. Below is a sample set of War Room rules.

War Room Rules

❑ Check in daily and attend 8am standup meeting
❑ Update status weekly or when tagged
❑ Do not erase or cover-up someone else's work
❑ Make effort to attend all team meetings
    ❑ >90%, buying the team breakfast for missing each additional meeting
❑ Create a schedule/calendar for room uses/activities
    ❑ Reserve work spaces
    ❑ Sign up meeting time slots
❑ No side meetings when a main team meeting is in progress
❑ Maintain confidentiality
    ❑ Notify the team when bring in visitors, including customer and management visits
    ❑ No private or confidential conversations in the room
    ❑ Checking with the individuals who are in the room working before holding meetings
❑ Use "Do Not Disturb" sign for critical meetings and individual work
    ❑ Respect the "Do Not Disturb" sign
❑ No personal phone call in the room
❑ Keep the room organized and clean

Example War Room Rules

Typically, a Project War Room should be big enough for the entire project team and co-locating all project members into the War Room is preferred. For big projects, a system of War Rooms can be setup with a War Room for each of the lowest level sub-teams and another War

Room for their team leads at a higher level. Some large projects may have several levels of War Rooms similar to the organization chart with the highest level War room called Project Central. Although it is not preferred, sometimes a project team must have members in different geographic locations. In this situation, a War Room in each location can be setup with remote control cameras installed so that members can see the activities in the other locations.

Managing Core Project Execution

After the physical environment is set up for the Core project teams, we shift focus to the project management approach. As mentioned in Chapter 3 when MBPM was introduced, Agile Project Management is more suitable for managing Core projects. In general, Agile methods create a fast tempo which is what we seek when executing Core projects. For instance, Scrum, the most popular Agile method, is executed in the cadence of regular Sprints and requires team members to attend daily Scrum meetings in person. If Scrum has been widely accepted and the teams are already used to the method, the company can just apply it to the Core projects. However, if Scrum has not been introduced into the organization, my recommendation is to adopt a simplified version of the Agile approach.

Many of the Scrum principles are useful for managing Core projects, however, there are too many rules, many of which are quite strict. Confining the Core project teams with rules and burdening them with administrative processes will negatively affect the velocity of project execution. We will not discuss the Agile rules and processes in details.

Instead, we will focus on the intention of simplifying these practices to increase efficiency when managing Core projects.

Scrum has some clearly defined roles: Manager, Scrum Master, Product Owner and team members. Each role also has well-defined responsibilities with both dos and don'ts. As mentioned earlier in this section, a dual-leadership model, with each team having a coach and a captain, is preferred in order to reduce administrative functions. The coach primarily serves as the manager role but also takes on some of the Scrum Master responsibilities, especially during the team forming and storming stages. Once the team reaches the norming stage, the team should be able to self-administrate its mode of work and meeting processes, making the facilitating role unnecessary and potentially counter-productive.

The Product Owner role in Scrum should be largely fulfilled by the team as a whole with the coach and the captain stepping in for high-level decisions or to resolve disagreements. The team should define work by initiating and prioritizing tasks together, preferably using WBuS. In Scrum, the Product Owner accepts and rejects work, which is similar to the final inspection step in most operations and is considered a waste under the Lean philosophy. The task owners should know when their task are complete. Work acceptance and quality should be determined by the receiving team members who conduct the subsequent tasks. The quality grading process discussed earlier serves the purpose of ensuring work is done well. The final gatekeeper for the entire project should belong to the entire team with the coach and the captain taking the lead roles. In short, we practice Agile in concept by trusting the team to define roles and responsibilities so that the team can be self-administrating without being regulated and monitored. Reducing overhead administrative functions helps to speed up the Core project execution.

Another rule that may need modification is the set cadence for the Sprints. While every Sprint having the same length instills discipline and simplifies administration, it also encourages procrastination. People who practice Scrum will most likely experience tasks being extended to the next Sprint. Core projects may still need to be separated into cycles similar to the Sprints, but the length of these cycles may vary based on the task schedule, preferably using Critical Chain Scheduling. The team should develop an overall cycle plan at the beginning project planning stage and finalize the length of the next cycle before exiting the current cycle. Essentially, near the end of each cycle, the team will get together to develop the Critical Chain Schedule for the next cycle and then determine the length of the cycle based on task durations and buffers. When the work is done early, the team moves on to the next cycle and does not need to wait until the end of the Sprint as in the case when strictly practicing Scrum.

The cadence of Core project execution is somewhat driven by using fixed cycles such as Sprints. It is more effective, however, to establish consistent activities with a set tempo. On a daily basis, the team should start with a morning standup meeting to quickly share the plan for the day, and in late afternoon another standup meeting is held to share progress. Attendance is openly tracked to create peer pressure under the cultural approach to managing performance. For each cycle, a client or management checkpoint should be held at the beginning to confirm objectives and then again at the end to review results. After the client or management checkpoint is completed at the end of a cycle, the team should spend time to reflect on the cycle execution and share learnings. While we can be flexible in applying the Agile rules related to roles and responsibilities as well as cycle lengths, trusting the team to execute with fewer constraints, we also need to utilize these standardized activities to create pressure and discipline to drive results.

Checklist for Managing Core Projects

❑ Define the Core project strategy
   ❑ Survival, growth, or both
❑ Select Core projects
❑ Set priority
❑ Communicate to the entire organization
❑ Form Core project teams
   ❑ Core project team members
   ❑ Core project leaders
❑ Move the team from forming to performing quickly
   ❑ Logistics and ground rules
   ❑ Tasks and schedules
   ❑ Quality assessment and grading measures
   ❑ Practices for accelerating Core project velocity
❑ Establish Project War Rooms
   ❑ Design and layout
   ❑ Setup
   ❑ Rules and operating practices
❑ Practice Agile with simplicity and flexibility
   ❑ Roles and responsibilities
   ❑ Cadence
   ❑ Standardized activities

Checklist for Managing Core Projects

A checklist for managing Core projects is provided above. As you should know by now, Core projects are the most important endeavors for a company and they must receive the highest priority and focus. Failure is not an option, so select the best candidates with the right personal characteristics to form action-oriented teams and provide them with the support structures necessary to enable highest possible quality and velocity. Agile Project Management practices are the most suitable for managing Core projects but apply them with simplicity and flexibility.

Managing Core Projects

Exercise Questions

E-mail your thoughts on the questions to MBPM.Innovation@gmail.com I will share my thoughts and answers to the questions.

Rules: (For details and reasons, please read Preface Page xvii)

1) One question at a time and state the Question # in the subject line of the email.

2) Provide a scanned copy of the book purchase receipt the first time you use an email to send in a question. This won't be necessary for future questions using the same email.

3) My response will be sent to you between 2-4 weeks after I receive your email.

Q9-1. A company's Core project selection is often based on its near-term objective, which can be survival, growth or both. What are the environmental factors that determine the chosen objective? What happens if the business environment is changing dynamically? How can a company be prepared to react if its business environment changes quickly?

Q9-2. A company should have a clear focus and avoid conducting too many Core projects. Do you think it is better to do one Core project at a time? What are the advantages and disadvantages of doing so?

Q9-3. We emphasize quality and velocity in Core project execution. To manage these effectively, we need indicators to measure the progress made in these two areas. These indicators must be able to show the status of the project in each area and enable benchmarking from one project to another. What should we do in each of these areas?

Q9-4. We discussed building the Core project team as a sports team. Do you think we should have backup players? Should we allocate extra resources to the Core project team?

Q9-5. Project War Rooms are recommended only for the Core project teams. How about the PF projects that are in the top-secret category? Should highly confidential projects also have dedicated rooms? What should we do?

Q9-6. We discussed how the Agile approach should be adapted for simplicity and flexibility. Should we train all Core project teams on Agile as a foundation before the adaptions are applied? What should we do to introduce the modified approach?

Managing Continuous Improvement Projects

Strategy for Participation and Engagement

Continuous Improvement (CI) projects are not only for efficiency improvements but imperative to setting up a culture of innovation. As mentioned in the introduction of MBPM in Chapter 3 and the discussion of corporate culture in Chapter 4, it is important to create an inclusive culture where everyone participates in projects. Innovation is not the job for employees in research labs and development groups alone. When all the employees in a company – operators, technicians, administrative and support staffs included – strive to exceed the expectations of their jobs, which is how innovation is defined in this book, significant peer pressure is created for engineers in the R&D departments to perform beyond expectations as well; this is the powerful aspect of a good corporate culture.

Everyone participating in projects means the company is doing numerous projects. What is the origin of these projects? Are there enough projects for all employees? More importantly, can we afford the resources to conduct these many projects? First, management should not be responsible for finding projects for employees. However, it is the job of management to encourage and challenge employees to pursue continuous improvements and initiate projects. Second, improving the current processes should not be seen as a burden but a mode of work embedded in daily routines. The continuous improvement effort increases the efficiency of the current processes and often free up resources for additional projects.

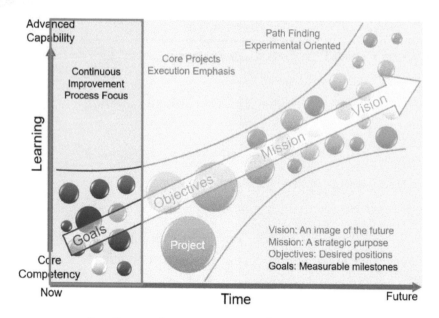

Continuous Improvement Projects in MBPM

In the T section of the last two chapters, the initiation of Path-finding (PF) and Core projects is discussed. Similarly, the initiation of CI projects is discussed in this section, but somewhat differently, here a bottom-up approach is more effective than a top-down mandate.

There are three directives to drive employee participation and engagement in CI projects. The first directive is utilizing the Lean philosophy, which focuses on eliminating waste in existing processes and systems. People who have implemented Lean would most likely advocate that Lean is not a program but a culture which involves everyone seeking to remove waste. In principle, Lean aligns with our intention to establish a culture with bottom-up initiatives, but the conventional approach of Lean introduction and implementation is not entirely appropriate for our purpose.

Many companies implement Lean to save time and costs as they emphasize the cultural aspect of Lean, communicating to employees that this is an ongoing effort. This strategy, however, goes against nature and will likely peter out gradually because returns are higher at the beginning when a new initiative is implemented but eventually level off. It is relatively easy to identify waste when Lean is first introduced but ultimately becomes much more difficult.

Management tends to measure Lean success by the amount of waste eliminated, or time and expense saved, which pushes employees to cut and cut until there is nothing to cut but corners. Just like a person pursuing weight loss, persistence is needed but the process cannot be endless. Even if management does not use these measures to manage employee performance, it would still be difficult for employees to continue expending effort when they see minimal benefit from their actions. In addition, requiring employees to do something endlessly is physiologically demotivating because they never see the end of the tunnel. It is wise to utilize Lean to get the bigger benefits and then shift the effort to other initiatives when the returns start diminishing. It is too idealistic to believe that setting up a Lean culture alone is good for a company in the long run. Lean does offer tools to help employees to identify inefficiencies in their work, which is useful since sometimes projects are initiated as a result. We are utilizing Lean to set up an autonomous culture of innovation, not to just have a culture of Lean.

To build a healthy body, losing fat is only part of the goal. Likewise, Lean needs be complemented with other efforts to achieve organizational health. This takes us to the second directive in driving CI projects, which is challenging employees with specific efficiency improvement goals and then allowing them to initiate projects towards meeting those goals. While Lean is like reducing fat when building a healthy body, this second directive is equivalent to strengthening muscles and mental health. The approach is to select key areas of the business, review the current situation along with relative metrics and indicators, and then set improvement goals to create value instead of just reducing inefficiencies. Meeting these goals may require the introduction of new tools, new processes and new systems that are typically filled with inefficiencies themselves, but this is fine since we can reintroduce Lean to clean up the waste after. These initiatives are done in cycles repeatedly, similar to the Tick-Tock Model used by Intel: one cycle is focus on architectural improvements while the next cycle is focus on process improvements.

Both of the above directives are from the company's perspective but an employees' perspective is also needed to drive CI projects effectively. As such, the third directive is to position CI projects as a part of employee career development. Individuals advance their careers by taking on learning opportunities and additional duties beyond the scope of their current jobs. Management should collaborate with HR to roll out career development training and workshops for all employees to inform them that seeking and participating in CI projects is a natural avenue for career development. Management should also offer programs and support to help employees engage in career development opportunities. When a majority of employees actively pursue career advancement, numerous projects will be initiated and executed by these self-motivated employees.

Participation and engagement of CI projects can be driven by a combination of the three strategic directives: Lean methodologies, improvement challenges and career development initiatives. The more people participate, the better the results will be. Inclusion is therefore the key. When there are many employees participating, the effort must be done in a coordinated fashion to yield the best result. Thus, the next section's discussion topic will be the inclusion and coordination of CI projects. The L section of this chapter will enumerate tactics for driving and managing CI project initiatives under the three strategic directives.

Managing Continuous Improvement Projects

Inclusion, Coordination and Prioritization

Building a culture of innovation is our goal and conducting projects is the vehicle for achieving this goal. We cannot just rely on the research and development groups to achieve this. The more people exhibiting similar behavior, the stronger the culture will be. Ideally, we want all employees participating in projects. To entice employees to join the effort, we need to educate everyone as to how innovation is not necessarily big and radical. In fact, most innovative successes are incremental and involve simply reapplying known knowledge to different areas. Therefore, we can categorize innovation into three types: radical, reapplied and incremental.

Our definition of innovation is to exceed the customer's expectation. As such, every employee can pursue a project aimed at exceeding the expectation of his or her customer in his or her work area. We can furthermore group innovations based on business area, such as strategic, business model, organizational, marketing, product, service, process and supply chain:

- Strategic innovation is related to the company's vision, mission and strategies, such as a new innovative direction, competitive positioning or growth plan.

- Business model innovation is about new ways of doing business, such as a unique profit model or a new approach to value creation.

- Organizational innovation is associated with human resource management, such as a new employee involvement practice, organizational structure or culture.

- Marketing innovation is about a new means of channeling the company's products and services to consumers, such as a new form of advertisement, differentiation or segmentation.

- Product innovation is about new functions and features in product designs. Service innovation is about new approaches in serving customers, extended care or new offerings.

- Process innovation is associated with new techniques and methods that improve the efficiency and quality of the processes used to make products or deliver services.

- Supply chain innovation is related to new methods to manage the flow of materials and supplies, such as outsourcing, offshoring, partnership, etc.

Below is a matrix showing examples of innovation by type and area.

Innovation by Type / Innovation by Area	Radical Innovation	Reapplied Innovation	Incremental Innovation
Strategic Innovation	Apple	Samsung	Xiaomi
Business Model Innovation	Airbnb	Uber	Lyft
Organizational Innovation	Google	Warby Parker	Facebook
Marketing Innovation	Pepsi	Nike	Starbucks
Product Innovation	Tesla	BYD	Nissan
Service Innovation	Amazon	Alibaba	Zappos
Process Innovation	Ford	Airbus	Toyota
Supply Chain Innovation	Dell	H&M	Cisco

Innovation Matrix with Examples

Some readers may feel that some of the examples in the matrix are not quite correct. This may be because there are many viewpoints and sub-categories within each area. For instance, in organizational innovation the examples are derived from a corporate culture perspective. If we are viewing from an organizational structure perspective, Zappos may be a good example for its success in using Holacracy instead of Google. When presenting the matrix to employees, invite them to come up with examples and explain why. Don't get too caught up in the details. The purpose of the matrix is to help initiate a discussion and demonstrate that innovation can be done in almost any area. For those who want to get to a more detailed level, the matrix can be expanded with additional business areas. Also, many companies, such as Apple, Google and Amazon, can serve as examples in many of the matrix's cells. That is because these companies have a culture of innovation, which is exactly what we want.

After seeing some great innovation examples, it is time to challenge employees to build this culture together so the company can be seen as an innovator in many business areas. The same matrix can be utilized again but with projects entered in the cells instead of companies.

Innovation by Type / Innovation by Area	Radical Innovation	Reapplied Innovation	Incremental Innovation
Strategic Innovation		Project A	Project B
Business Model Innovation		Project C	Project D
Organizational Innovation	Project E	Project F, G	Project H, I
Marketing Innovation		Project J	
Product Innovation	Project K	Project L, M	Project N, O
Service Innovation		Project P, Q	Project R
Process Innovation	Project S	Project T	Project U, V, W
Supply Chain Innovation		Project X	Project Y, Z

Example Innovation Matrix with Projects

The new matrix clearly communicates the connection between projects and areas of innovation so that effort can be spent in a coordinated fashion. It is not necessary to fill up the entire matrix with projects in all cells but there should be projects in each area so everyone can participate in at least one project. Most CI projects should target incremental innovation and reapplied innovation but on some occasions, we can shoot for radical innovation. Again, not all CI projects need to reach a successful result but everyone must try. The key is to cover all areas and provide visibility for projects to all employees, encouraging project initiation and participation. The matrix may be done at the corporate level or at the division level for large companies.

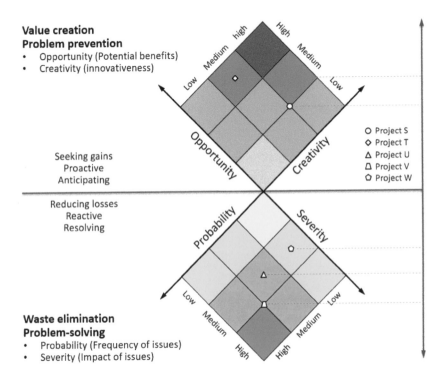

CI Project Prioritization Profiling

If we successfully achieve inclusion in project participation as a means for creating a culture of innovation, we will find ourselves conducting a great number of projects. In addition to coordinating effort among different areas, we also need to prioritize effort when there are many projects proposed in a particular area, exceeding the availability of resources. The above illustration demonstrates the method used to prioritize CI projects. The concept is similar to the profiling system used in WBuS risk handling shown in Chapter 6 Page L6-13.

CI projects are separated into two categories: waste elimination and value creation. Using the fire department as an example, the first category of projects are improvements to the firefighting operation, which may include shortening response processes, keeping equipment reliable and improving the skills and fitness of the personnel, whereas the second category of projects are community outreach programs, which may include educating the public on fire prevention and performing targeted inspections of old buildings.

Projects in the waste elimination category aim at reducing deficiencies in the current system, which typically means identifying inefficiencies and problems and then resolving them. Prioritization of these projects is based on a combination of ratings concerning the frequency of the issues and the size of their impact, which essentially means solving the biggest problem first. Projects in the value creation category are proactively changing the current system to increase benefits or prevent undesired issues from reoccurring. In order to maximize the gains, they are rated based on a combination of potential benefits and innovativeness.

As an alternative to the profiling illustration, a simple Excel worksheet can be set up with formulas to calculate the priority of each project. The table shown on the next page has the same projects with the same rating assessments as shown in the previous illustration. The score for a project is calculated by "=Probability*Sin(45°)+Severity*Cos(45°)." Should two projects arrive at the same score, a clear order can be achieved by modifying the ratings or adding a weighted factor in the total score formula. When two projects under the same category have the same score, we can reassess either or both ratings. When projects in different categories have the same score, we can give priority to one category over the other. In the example shown in the previous illustration, projects in waste elimination have a higher priority than those in value creation. The total score formula is thus "=Score1*1.01+Score2" where Score1 is the score of the waste elimination project and Score2 is the score of the value creation project.

| Projects | Waste Elimination | | | Value Creation | | | Total | |
	Probability	Severity	Score	Opportunity	Creativity	Score	Score	Priority
Project S			0.00	2	4	4.24	4.24	4
Project T			0.00	5	3	5.66	5.66	2
Project U	3	3	4.24			0.00	4.26	3
Project V	4	4	5.66			0.00	5.69	1
Project W	1	3	2.83			0.00	2.84	5

Scale: 0-6 with 6 the highest

Example CI Project Prioritization Worksheet

A company can set a quarterly budget for CI projects and then conduct projects based on the priority scores until the entire budget is allocated. Some CI projects, however, actually save money for the company and, as a consequence, the savings can be added to the CI project budget to fund additional projects. Management should review the budget regularly and enable a mechanism for appeals in order to ensure that a reasonable volume of good CI projects are being executed.

Having discussed inclusion and coordination of our CI project effort by mapping and prioritizing the projects into business areas, we are moving on to planning and executing these CI projects, which is the main discussion topic of the next section.

Managing Continuous Improvement Projects

Autonomous Culture of Innovation

Given that CI projects are typically smaller in scope, large in quantity and executed by many people, they can and should be run in a structured manner. The key principle of Traditional Project Management – plan the work and work the plan – is mostly suitable. Since we are aiming for an autonomous culture of having everyone in the company initiate and manage projects, we need to train all employees on project management, but it is unrealistic to have everyone learn the 49-process project management procedure developed by the Project Management Institute (PMI). Therefore, we need to simplify the project processes so all employees, including technicians, operators and administrators, can easily initiate, plan and execute projects. For that, we utilize Lean to eliminate the non-value-added activities in the Traditional Project Management processes.

CI Project Templates and Processes

Lean offers many forms and templates such as the Knowledge Brief (K-Brief), A3 and Waste Elimination Worksheet to help start the process. These templates are typically one page and incorporates problem-solving models such as PDCA (Plan-Do-Check-Act) or LAMDA (Look-Analysis-Model-Discuss-Act). A couple of Lean example templates are shown on the next page. We can simply borrow the Lean practice of utilizing a one-page form to help start the CI project effort and guide the process.

Title:			Date: _____
Author(s):	Sponsor:		Revision: _____

Background:	**PLAN**	Counter Measures	**DO**
		What Who When	
Current State:			
Goals:			
Root Cause Analysis:		Results:	**CHECK**
Future State:		Follow-up Actions:	**ACT**

Example Lean A3 Template Utilized PDCA

Type of Waste: ☐ Defects   ☐ Waiting   ☐ Inventory   ☐ Motion
☐ Transportation   ☐ Overproduction   ☐ Over-processing

Observation:

Current State   Date: _____   Improved State   Target Date: _____

Measurable Benefits: ☐ Performance   ☐ Quality   ☐ Cost
Explanation:

Originator: Team Members:

Example Lean Waste Elimination Worksheet

However, we should modify the template to include innovation as well as both the value creation and waste elimination categories. A suggested CI Project Initiation Form is shown on the next page.

Continuous Improvement Project Initiation Form

Project Title:				Date:
Value Creation:	☐ Option/Offering ☐ User Experience	☐ Knowledge/Skill ☐ Safety/Ergo	☐ Technology/Techquie ☐ Corporate Image	
Waste Elimination:	☐ Defects ☐ Transportation	☐ Waiting ☐ Overproduction	☐ Inventory ☐ Over-processing	☐ Motion

Past & Present:

Future State:

Phase	Status	Date

Analysis:

Value Creation:	Opportunity Score: ___
	Creativity Score: ___
Waste Elimination:	Probability Score: ___
	Severity Score: ___
Priority:	Total Score: ___

Result & Learning:

Targeted Innovation:	Strategy	Business Model	Orgnization	Marketing	Product	Process	Supply Chain
Radical	☐	☐	☐	☐	☐	☐	☐
Reapplied	☐	☐	☐	☐	☐	☐	☐
Incremental	☐	☐	☐	☐	☐	☐	☐

Originator:	Team Members:

CI Project Initiation Form

The CI Project Initiation Form should be available in printed copies for easy access and posting. For people who prefer to fill out information electronically, a soft copy file could be made available on a share drive. After filling out the information, however, the form still needs to be printed. We need to display the physical form to drive participation and manage the execution process. We will further discuss the execution process management after explaining the use of the CI Project Initiation Form.

When an employee has an idea, he or she can use this one-page form to start the process. At the beginning, we should not require the employee to fill out all the information in detail. The printed form will be used for further collaboration with the team and it will be a living document throughout the entire lifecycle of the project. It was discussed that we separate CI projects into two categories: value creation and waste elimination.

When proposing a value creation project, an employee can further select one or more of the values that the project is anticipated to bring, like introducing a new option to existing product lines or offering additional services, creating new knowledge aligned with our future vision, developing a new technique or model that can be applied to improve existing processes, enhancing the user experience in using our products and services, preventing safety and ergonomic issues, or lastly promoting the corporate image that will produce long-term benefits for the company.

Waste elimination projects are under the usual pursuit of Lean by examining the waste in existing processes. Employees can select one or more of the seven typical types of waste observed and targeted to remove. The details of each waste category will not be explained in this book; more details can be found in Lean literatures.

The "Past and Present" section is for the employee to provide a background for the project which typically consists of a history and the current state of a situation. This is a section for facts and data. Evaluations of the situation should be stated in the "Analysis" section. For value creation projects, the "Analysis" section should present the clear link between the proposed actions and the created values, essentially providing a justification for the project. For waste elimination projects, this section presents the root cause of the issue triggered by the waste. It should also include scores for prioritization as well as innovation assessment. The details of the scoring method and innovation matrix were discussed in the previous section. The CI project originator may provide preliminary project scores and an innovation assessment but it is best to do these by collaborating with the team.

The "Future State" section is for proposing high-level phases needed to create values or remove waste. It is not intended for documenting a detail project plan, which should be done separately following the task and schedule management techniques presented in Chapter 6 and 7. This section also provides space for status updates as the project is being executed. Don't worry about limited space as we can put additional Post-it notes on top of the form if needed. The "Result and Learning" section is also for future use and, likewise, additional notes can be attached to the form.

After the forms are generated, they should not be kept as records in a pile or files in computers. These forms need to be displayed so employees can see what projects are being done. One of the purposes of the posting is to highlight to everyone who are the active participants, which utilizes the cultural approach to management that we have discussed many times previously. Second, employees can see the projects in action, which encourages participation on projects of interest and reduces duplicated efforts. Third, the display promotes collaboration by making it easy for employees to comment and contribute ideas by writing notes on the forms or attaching Post-it notes to them. Lastly, the displayed forms are utilized to enhance execution management by progressing them from phase to phase with updates.

To enable these updates, we need to design the display board for these forms. I call it Project Lifecycle Display as the project execution process flow are shown and enable the forms to be displayed in each project life cycle phases. An example setup is shown below.

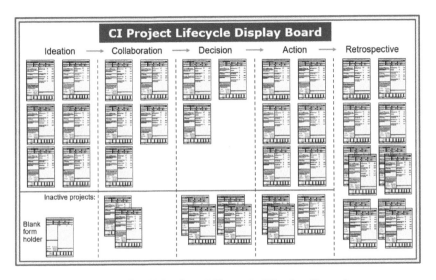

Example CI Project Lifecycle Display Board

There are many options for the process flow. The first option is to use PDCA with areas on the board for Plan, Do, Check and Act. The next is to use the typical project life cycle model in Traditional Project Management: Initiation, Definition, Execution and Implementation. Custom flow can also be used based on the nature of the business and operation, such as Approval, Experiment, Prototype, Test and Follow-up. The example illustration uses the information management flow discussed in Chapter 5, which aligns with the information management tools described in that chapter (Page M5-6). This setup provides guidance to project teams on what tools to use in each phase.

The bottom part of the board has an area for inactive projects. These projects are 1) below the priority line due to resource constraints, 2) temporarily on hold due to incomplete data or pending additional verification or 3) failing to yield desirable results since CI projects are not required to reach the end.

Failure is tolerated because if all CI projects are completed successfully then employees are not being challenged enough. When deciding to put a project into an inactive state, the form can be simply moved below the line at the phase that was executed. Please note that we should not have inactive projects below the Ideation and Retrospective phases. All ideas are welcomed and projects should only be moved to inactive after they are shared with the team in the Collaboration phase. We typically use the area below the Ideation phase and place a plastic holder for blank project initiation forms for easy access and filling. This area can also be used to hold blank Post-it notes, user guides for tools and instructions of this CI project management process.

Logistically, the board should have places to hold pens, markers, color highlighters and magnets or push pins for attaching the forms depending on the type of board used. There should be additional boards nearby for the team to work on project prioritization, task management and schedule. The Innovation Matrix and Prioritization Worksheet should also be displayed nearby to provide an overview of all projects. For large groups and departments with many employees, we can setup one board for each phase. These boards must be placed in high traffic areas to allow maximum viewing and interaction. For an even bigger participating group, we can setup one conference room for each phase with multiple boards in the room. For better organization, each board can be dedicated to a further subdivided subject or area.

CI Project Engagement Training

After the forms and templates are developed, the next step is to train all employees to be capable of initiating and executing CI projects autonomously. A recommended training curriculum is shown on the next page. It consists of five main modules and takes a full 8-hour work day to complete. Of course, it can also be done in two half-day sections or five short sections by each module.

The first three modules are designed to build the foundation for an autonomous culture and the last two modules are the practical guide for initiating and executing CI projects. There is also an introduction at the beginning to connect the company's vision and strategy to the CI project effort, so employees have a better understanding of why these initiatives are taking place.

Module 1 is about innovation. The main purpose is to convey that innovation is simply about exceeding the expectation of users or management. We challenge employees to do great things that are beyond the call of duty. Innovation is not necessarily radical and there are three types of innovation that can be done in many business areas. Present the Innovation Matrix and invite employees to cite examples then shift the focus to the company and ask the employees to brainstorm (or Brainwriting and Brainswarming) possible innovations to put on the Innovation Matrix. Companies that already have a culture of innovation may reduce the content of this module or even skip this module.

CI Project Engagement Training Curriculum				
Module	Description	Material Source	Estimate Time	Schedule
Introduction	Introduction - Link the company vision & strategy to project effort	Company vision & strategy	15 min.	8:00-8:15am
1 Innovation	1.1 Definition of Innovation - Exceeding expectation	MBPM Chapter 1 Page T1-1 to T1-3	1 hour	8:15-9:15am
	1.2 Innovation by area & by type - Interactive: ask students for examples	MBPM Chapter 10 Page M10-1 to M10-3		
	1.3 Innovation Matrix for the company - Interactive: Brainstorm possible projects	MBPM Chapter 10 Page M10-3 to M10-4		
	Break		10 min.	9:15-9:25am
2 LEAN	2.1 LEAN Overview	LEAN literatures (Limited discussion in MBPM Chapter 10 Page T10-2 to T10-4, L10-1 to L10-3 and L10-11 to L10-12)	1 hour 15 min.	9:25-10:40am
	2.2 Key LEAN Tools			
	2.2.1 Waste Elimination & 7 Wastes			
	2.2.2 PDCA			
	2.2.3 LAMDA			
	2.2.4 VSM (Value Stream Mapping)			
	Break		10 min.	10:40-10:50am
3 Career Development	3.1 Career development overview - TOP model	MBPM Chapter 8 Page T8-2 to T8-3	1 hour	10:50-11:50am
	3.2 Career development Planning	MBPM Chapter 3 Page L3-1 to L3-4		
	3.3 Career development & CI projects participation - learning opportunities & growth in scope	MBPM Chapter 10 Page T10-5		
	Lunch		1 hour	11:50-12:50pm
4 Processes & Templates	4.1 CI Project Initiation Form	MBPM Chapter 10 Page L10-3 to L10-6	1 hour	12:50-1:50pm
	4.2 CI Project Prioritization Profiling & Worksheet	MBPM Chapter 10 Page M10-5 to M10-7		
	4.3 CI Project Process Management - CI Project Lifecycle Display Board	MBPM Chapter 10 Page L10-6 to L10-8		
	Break		10 min.	1:50-2:00pm
	4.4 CI project initiation trial - Interactive: ask participants to practice filling out an initiation form for an actual project	Do this in small team, help and share among members	45 min.	2:00-2:45pm
5 Models & Tools	5.1 Information management models & tools	MBPM Chapter 5 Page M5-6 to M5-7	30 min.	2:45-3:15pm
	5.1.1 Brainwriting & Brainswarming	Page L5-1 to L5-11		
	5.1.2 Six-Step Decision-making Process & RAPID Decision-making Model	Page L5-22 to L5-28		
	5.1.3 Convergent Problem-solving	Page L5-28 to L5-32		
	Break		10 min.	3:15-3:25pm
	5.2 Task management - WBuS overview	MBPM Chapter 6 Page M6-4, L6-1 to L6-12	40 min.	3:25-4:05pm
	5.3 Schedule Management - Critical Chain scheduling and tracking overview	MBPM Chapter 7 Page L7-1 to L7-22	40 min.	4:05-4:45pm
Summary	Wrap-up		15 min.	4:45-5:00pm

Recommended CI Project Engagement Training Curriculum

Module 2 is about Lean methodologies, which was originally practiced by manufacturing companies but many companies have since extended it to product development. Even if the company has introduced and practiced Lean previously, a refreshed training focusing on a few key practices is recommended. First, training on Lean waste elimination is extremely important as it is one of the two CI project categories and typically the largest source of CI projects. Employees need to understand the seven wastes and train their eyes to recognize these wastes in their daily job activities. PDCA and LAMDA are useful tools for guiding employees to identify and solve issues. Value Stream Mapping (VSM) is another useful Lean tool to help identify and remove waste, so I will explain in further detail.

VSM is a Lean technique used to analyze a work process flow. Activities in the process flow are categorized as either value-adding, necessary waste or waste. The goal is to build a future state process with the least amount of waste. This is typically done in a team environment using Post-it notes to represent activities and then mapping the current process state on a board. Data for each activity, such as process time, inventory, materials and quality requirements are also captured. Then each activity is analyzed and categorized using colored stickers on the Post-it notes with green for value-adding, red for waste and yellow for necessary waste. Finally, redesign the process flow by removing or reducing waste and necessary waste.

Unconventionally, I apply the VSM concept in career mentoring to help my mentees manage their time. The most common excuse I hear in career mentoring sessions is that "I am very busy and don't have time." I respond by showing the following time calculation:

- There are 168 hours in a week.

- We typically work five 8-hour days but let's add an extra day per week. That is 48 hours and we have 120 hours left.

- We need to sleep and the recommended sleep time for adults is 7 hours per day (6 hours may already be appropriate typically) so the total is 49 hours. We have 71 hours left.

- We eat 3 meals a day and let's use half an hour per meal average. Dinners may take longer but breakfasts are typically shorter. The total eating time is 10.5 hours. We have 60.5 hours left.

- For personal hygiene, let's put 1 hour per day so we have 53.5 hours left.

- We want to stay healthy so we need to exercise. The Department of Health and Human Services recommends 30 minutes exercise per day for healthy adults. So we deduct 3.5 hours.

- We still have 50 hours left, which is an amount greater than any of the activities we have listed.

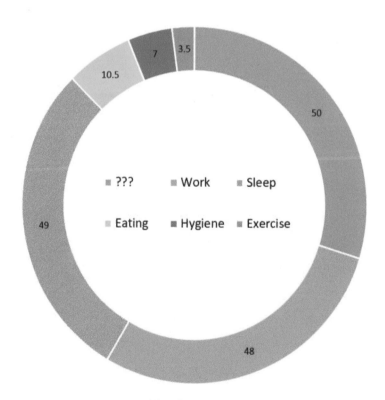

Weekly Time Breakdown

Some people may continue to list other activities such as family time, socializing with friends, commute, etc. At the end, there will still be a considerably large amount of time that is not accounted for.

In the book *Outliers*, Malcolm Gladwell repeatedly mentioned the "10,000 Hour Rule," which is the time takes to become an expert in a field. That is about 38.5 hours per week for five years, which is the typical duration for a doctorate program. I had started working full time before I decided to pursue a doctorate degree. At the time, I did the above calculation and convinced myself that it was doable. So I did it, and not just once. I earned doctorate degrees in engineering and business and received three doctorate degree certificates. The most valuable lesson I learned was how to make time through this process. Time, in my opinion, is the most limited resource that we have as we will never gain it back once it passes.

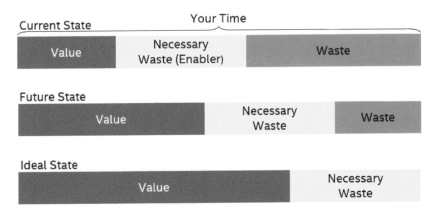

Apply VSM in Time Management

I encourage you to do the calculation and understand where your time goes, but if you want to make time, the next step is to pick a typical week and apply VSM. First, list all the typical activities and their durations. Next categorize the activities. Working time is green as you earn a salary. Sleep time is yellow but only up to the recommended hours and any additional sleep time is red. The same logic applies to eating, hygiene and exercise. If you feel that certain family time is very important, put it as green. Once this is done, you will know where your time is wasted and can build a new schedule to better utilize your time.

The purpose of doing this is not to schedule your life like a machine so that you don't have any fun. The purpose is to make you conscious of your time so at any moment, you know if what you are doing green, red or yellow. If you are doing red, try to hurry up and shift to yellow or green. It also allows you to plan your time better by combining green activities with red or yellow activities, such as allocating some meal time to socializing with friends. When you have high priority activities, you will also know where to make time available. This time management method worked well for me and enabled me to achieve some substantial accomplishments. I am relatively young but most people think that I am much older based on my biography – working full time in high-tech for over 30 years, teaching part-time at universities for almost 20 years, earning multiple doctorate degrees, writing several books, having a family and raising two children, in addition to engaging in many hobbies and volunteering activities. This time management method enables me to live my life the fullest.

This unconventional application of VSM in time management also applies to my time at work, which is already value-adding activity for me as I earn a salary in return, but not all activities that I do at work create value for my company. Using VSM, a typical work day can be analyzed and waste can be identified and removed as well. Clearly, VSM can be used for personal time management in addition to the typical work process management. Learning the VSM tool benefits both the company and the employee personally.

Module 3 is about employee career development and it consists of the TOP model, career planning, and the connection between career development and projects. Helping employees find growth opportunities not only demonstrates that we care about employees' personal development but also promotes a learning organization. An organization grows when its employees grow. Career development has been mentioned many times in this book since it has a significant connection to building the culture of innovation through projects. Projects are great opportunities for personal development and due to the resemblance of a career plan and MBPM, learning to plan a career helps individuals understand the importance of mapping projects to an organization. Educating employees on career planning increases employees' desire to initiate and participate in projects.

Module 4 is about the processes and templates used to manage CI projects. We need to train employees on how to fill out the CI Project Invitation Form, how to prioritize CI projects and how to use the Project Lifecycle Display Board. This module also includes an interactive session where participants, working in small teams, are asked to fill out a CI Project Initiate Form to propose a project that can be implemented to improve their current jobs. After projects are initiated and posted on the Project Lifecycle Display Board, ask each team to share and encourage other teams to provide input. The sharing not only helps students learn from each other but also starts the habit of sharing and collaborating, which should be continued after the training session.

Module 5 is about the models and tools that employees can use to plan and execute the projects that they initiate. My recommendation is to design content based on the Project Management model proposed in Chapter 5 (Page T5-1): managing information, tasks and time. Alternatively, this module can be customized based on the project management approach to which the company is accustomed, whether it is Traditional Project Management or Agile, but it is not for training project managers so the content should be simplified for the general population. The purpose of this module is to introduce the models and tools, not to teach these models and tools in extended detail.

We want the participants to learn by practicing and sharing their experiences while executing their projects. A learning organization is not built by conducting hours of class room training, rather much of the learning is achieved through actual practice. Also, we should not require employees to practice certain models and tools in certain ways. They should have freedom to choose and practice these models and tools in the ways that work for them and their situations. While practicing the learning, employees can propose value creation or waste elimination projects to improve the CI project management process, models and tools. Training sets the foundation and the actual CI project execution is where the real learning occurs and the culture is built.

Tactics in Driving Active Participation

A culture is built on openness, sharing, collaborating and inclusion. Tactics used in CI project execution must therefore encourage participation. First, visual displays should be set up to highlight key wins and gaps. Displaying key wins provides examples to others and inspires them to do the same. Showing gaps, especially participation gaps, creates peer pressure on the underperforming areas and groups to encourage increasing effort. There should be a central location for all high-level CI project displays, which includes the Project Lifecycle Display Board, the Innovation Matrix, the Prioritization Worksheet, key wins and gaps, as well as participation indicators. There should also be displays at the location where a project is being executed to show the detail of that project. For instance, if there is a project to improve the output of a station in a lab, there should be a small poster or board displayed at the location as we want to encourage action at the place where waste and values are identified. These displays must be reviewed and updated regularly; otherwise, they are merely decorations and eventually people will not pay attention to them anymore.

The second tactic in encouraging participation is therefore to schedule regular forums in front of the displays. The first forum is daily reflections, which is a short standup meeting in front of the local project displays to share the actions taken as well as plans and key learnings. All employees in the area should participate, even those who are not working on the project. Everyone is expected to provide feedback and learn from others. Inclusion is very important to setting up an autonomous culture.

These reflection sessions must be short, focused and productive. We do not want them to burden employees, so if daily meetings are too frequent and people become exhausted, they can be held every other day or twice a week. We also understand that there are some naturally shy people who may not be comfortable speaking in front of a group. We encourage them to write on the displays or on the Post-it notes.

This is one of the main reasons that we must setup the displays. These written inputs can be provided throughout the day and at the time of the reflection sessions a designated person in the team can read those notes and facilitate the discussion.

The second forum is an integration events, which is a longer meeting occurring at the central location of the CI project displays. The required attendees are project owners and key participants, but everyone should be invited to participate and observe. If the size of the organization is too large for conducting an effective meeting, these events can be separated into areas and then have an overall event with the area owners. These integration events should also occur regularly on a set schedule, typically weekly or bi-weekly. The purpose is to provide coordination and prioritization for all CI projects in order to avoid duplicating efforts, share learnings and create visibility to drive the culture of innovation.

In these regular forums, managers should not take a leading or chairing role. The role of management in the CI project execution is mainly mentoring and coaching. At the beginning, managers may participate in these forums and observe whether the process is working well. Ultimately, however, these forums should be conducted autonomously by employees without managers present.

The chair position for the forums may be rotated among the project leaders or event participants, providing inclusive opportunities to whoever has the desire to lead. Managers must make themselves available and approachable to mentor and coach the team leaders and members. Occasionally, managers may join the meetings to observe or seek feedback from key participants, and based on observation and feedback, they may openly discuss the findings with the team. I highly recommend that managers go through mentoring and coaching training to gain these skills. Great mentors and coaches help people see where they are and where they want to go. Typically, this is done through asking questions to guide people to find the right path and right solutions instead of providing assessments or directives.

We have discussed the use of rewards and incentives previously and stated that they should not be used based on project results since higher rewards lead to lower performance for non-mechanistic tasks. Recognitions given objectively based on measurable terms to encourage certain behaviors, however, are quite effective, especially in the beginning when people lack internal motivation to try. This is like getting children to learn a musical instrument or a sport. At the beginning, it takes a lot of repetitive practice, which is quite mechanical and boring. Providing incentives helps them get through this period. After they are good at the instrument or sport, the intrinsic motivations from enjoyment and the pride of winning and advancement will take over to keep them going.

We can therefore use rewards and recognitions to promote participation at the beginning of the implementation, such as incentives based on number of CI Project Initiation Forms proposed. After a majority of employees are engaged and getting good at the process, the CI project effort can then be maintained by intrinsic motivations and peer pressure from the culture. Occasional incentives can be used to maintain the momentum and to encourage new employees to join the effort. In addition, employees should be able to recognize other employees as peer-to-peer recognition strengthens the culture.

Our goal is to have everyone in the company participate in projects. The population of employees involved in CI projects is larger than employees engaged in PF projects and Core projects. Simplifying the initiation and administration of CI projects not only encourages more employees to participate but also creates greater efficiency collectively. We should also challenge employees to improve the way CI projects are managed. The following checklist serves as a baseline guide for managing CI projects. I am certain that it will evolve and be refined to a more efficient and a better fit to the organization through the collective input and experimentation of the employees in the company.

Checklist for Managing CI Projects

- ❑ Define participation and engagement strategy
    - ❑ Lean
    - ❑ Efficiency improvement
    - ❑ Career development
- ❑ Develop processes and templates
    - ❑ Innovation Matrix
    - ❑ Prioritization Worksheet
    - ❑ Project Initiation Form
    - ❑ Project Lifecycle Display
- ❑ Train employees
    - ❑ Innovation
    - ❑ Lean
    - ❑ Career development
    - ❑ Processes and templates
    - ❑ Tools for information, task and schedule management
- ❑ Apply CI project execution tactics
    - ❑ Visual displays
    - ❑ Daily reflections
    - ❑ Integration events
    - ❑ Mentoring and coaching
    - ❑ Reward and recognition

Checklist for Managing Continuous Improvement Projects

Chapter 10

Managing Continuous Improvement Projects

Exercise Questions

E-mail your thoughts on the questions to MBPM.Innovation@gmail.com I will share my thoughts and answers to the questions.

Rules: (For details and reasons, please read Preface Page xvii)

1) One question at a time and state the Question # in the subject line of the email.

2) Provide a scanned copy of the book purchase receipt the first time you use an email to send in a question. This won't be necessary for future questions using the same email.

3) My response will be sent to you between 2-4 weeks after I receive your email.

Q10-1. We discussed that there are three strategic angles to drive the CI project effort: Lean, improvement challenges and career development initiatives. Are there any additional ways to drive the CI project effort?

Q10-2. We want inclusion in project participation to create a corporate culture of innovation. Should employees in PF project teams and Core project teams be required to participate in CI projects as well?

Q10-3. We separate CI projects into two categories: value creation and waste elimination. Should we have a balance of quantity of projects in each area? That means we should not have only value creation projects or waste elimination projects. If we need to have both, should we prioritize the projects separately in each category?

Q10-4. Should we customize CI Project Initiation Forms and Project Lifecycle Display Boards for different business areas or use a standard for the entire company? What are the advantages and disadvantage of the options?

Q10-5. Who are the best trainers to carry out the CI project training sessions? Should they be managers, outside consultants or just regular employees? How should we ensure the training is conducted consistently throughout the entire organization?

Q10-6. Managers should not micromanage and direct CI project execution. In additional to mentoring and coaching, what can managers help to support the CI project effort?

Chapter 11
Epilogue

Overview of This Book

This book presents the concept of Management by Project Mapping (MBPM), aiming at continuous innovation through a new strategic view of project management along with a series of practical tools and concepts.

Key theme of each chapter:

Chapter 1:	Innovation
Chapter 2:	Project Management
Chapter 3:	Strategic Planning
Chapter 4:	People
Chapter 5:	Information
Chapter 6:	Task
Chapter 7:	Time
Chapter 8:	Exploration
Chapter 9:	Execution
Chapter 10:	Efficiency

Main focus of each section in the chapters:

T Section:	What – the concept
M Section:	Why – the explanation
L Section:	How – the practice
Q Section:	So – what do you think?

Infographic for this book:

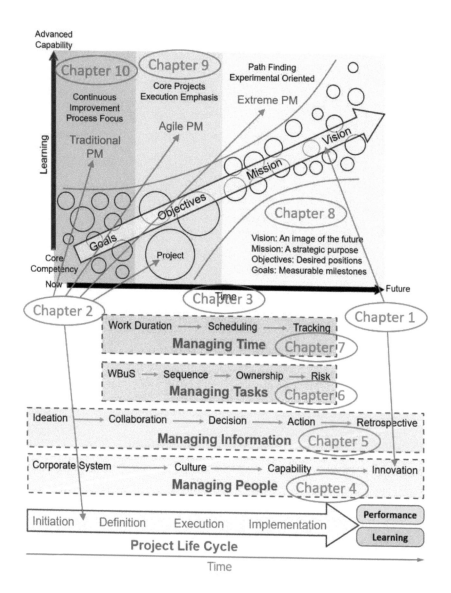

Chapter 11

Epilogue

Abridgment of This Book

Chapter 1: Corporate Innovation

Theme: Innovation

<u>T Section</u>

Innovation is achieving breakthrough results that exceed customer and market expectations.

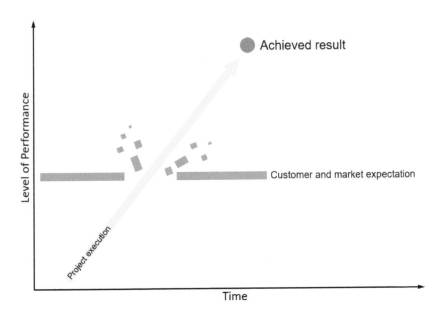

<u>M Section</u>

Corporate innovation can be measured differently and management needs to do more than just ask employees to innovate.

L Section

The journey of innovation is filled with challenges and requires a good vision and proper strategies for long-term results. It may start with a leader but needs to transition to a corporate culture-driven approach to sustain continuous innovation.

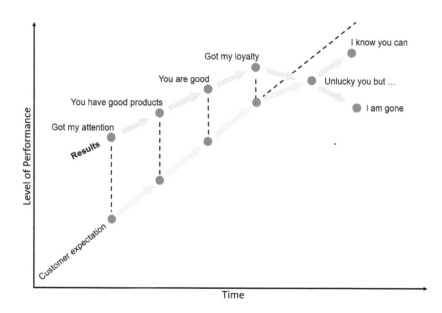

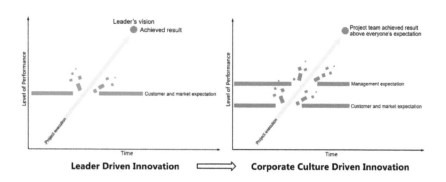

Leader Driven Innovation ⇨ Corporate Culture Driven Innovation

Chapter 2: The Project Management Field

Theme: Project Management

<u>T Section</u>

The target of project management is innovation.

<u>M Section</u>

Project management is a strategic means to enable a company to innovate towards its vision. Project failures are acceptable and project management should not be a control mechanism for achieving higher project success rate.

<u>L Section</u>

There are three general project management approaches: Traditional, Agile and Extreme. They are situational and the "one method fits all" approach will not work.

Chapter 3: Management by Project Mapping
Theme: Strategic Planning

<u>T Section</u>

MBPM is about using project mapping at the strategic level to transform a corporation's system, culture and capability to establish a foundation for continuous innovation.

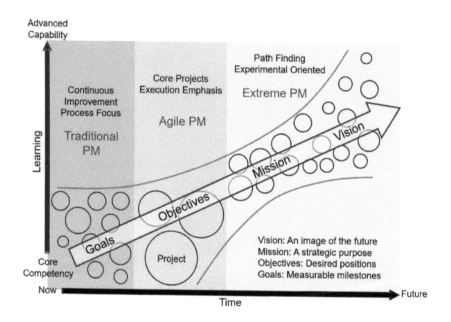

<u>M Section</u>

The key strategic elements of MBPM are 1) strategically utilizing projects to build a foundation for sustainable innovation, 2) mapping projects into three categories with alignment to corporate development, 3) using the right Project Management approaches for the right projects, and 4) promoting an all-inclusive innovative culture with projects for every employee.

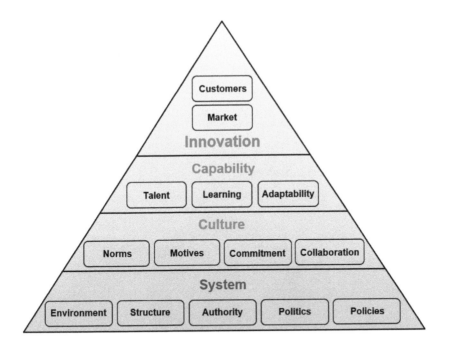

L Section

Projects are categorized into the areas of path-finding, core and continuous improvement with distinctive purposes, and they will need to be managed differently.

	CI Projects	Core Projects	PF Projects
Focus	Efficiency	Execution	Exploration
Timeline	Short-term	Mid-term	Long-term
Target	Productivity	Desired Position	Vision & Mission
Size	Small to Medium	Large	Small to Medium
Quantity	Many	A few	Many
Failure Tolerance	High	Low	High
Strategic Importance	Corporate Culture	Business Results	Capability Leadership
PM Approach	Modified Traditional PM	Modified Agile PM	Extreme PM
Executing Resource	Regular Employees	Development Teams	Research Labs

Chapter 4: Building the Foundations for Innovation

Theme: People

<u>T Section</u>

A corporation should build the foundation one layer at a time, starting from the corporate system to the culture then to the capability in order to achieve sustainable innovation.

<u>M Section</u>

The key ingredient for innovation is capable and motivated employees who can only be attracted and kept with the right corporate culture.

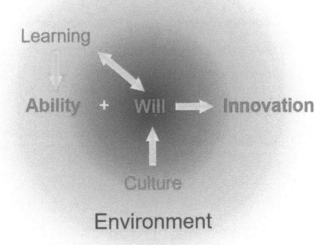

An organization can utilize project mapping and project management to build the foundation for continuous innovation. The foundation is composed of an agile, balanced and open corporate system, a committed and collaborative corporate culture and a capable organization with desires for challenge and mastery.

Chapter 5: Project Information Management

Theme: Information

<u>T Section</u>

Project information management system goes beyond the use of computerized tools and includes managing all activities in creating, organizing, processing and communicating information.

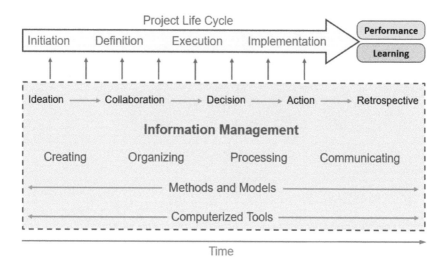

<u>M Section</u>

Managing project information covers the entire project lifecycle from ideation to retrospective with methods and models that are situational and should be flexibly applied with the least amount of bureaucracy and administrative effort.

Project Life Cycle	Initiation	Definition	Execution	Implementation	Performance / Learning

Methods / Models	Ideation	Collaboration	Decisions	Actions	Retrospective
Brainwriting	✓	✓			
Brainswarming	✓	✓			
CREATOR Meeting	✓	✓	✓	✓	✓
SMART Reporting	✓	✓	✓	✓	✓
Micromessaging		✓			
6-Step Decision Making Process			✓	✓	
RAPID Decision Making Model		✓	✓	✓	✓
Convergent Problem Solving		✓	✓	✓	
Project War Room		✓		✓	

L Section

Tools and models are situational. Learn by practicing.

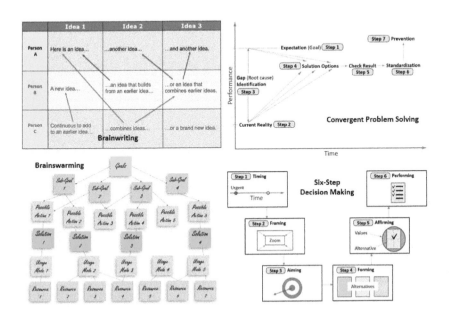

Chapter 6: The New WBS for Task Management

Theme: Task

<u>T Section</u>

Work Buildup Structure (WBuS) incorporates innovation into the task generation process, which is a better alternative to the traditional Work Breakdown Structure (WBS).

<u>M Section</u>

WBS is a top-down approach and WBuS is a bottom up approach. The key characteristics of WBuS, which are not presented in traditional WBS, are: 1) the team must work with customers to identify their expectations, 2) build a vision as well as tasks that lead to exceeding those expectations and 3) distinguish risk containment and risk-taking tactics based on tasks.

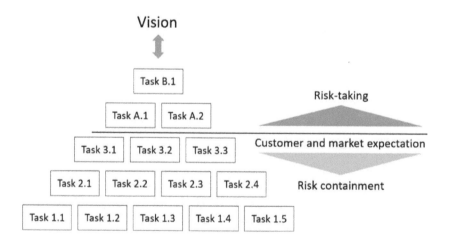

<u>L Section</u>

Developing task ownership and sequence map is a team process.

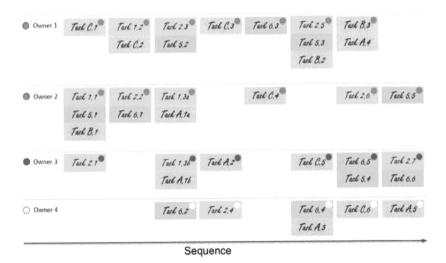

Sequence

WBuS Risk-handling methodology extends the traditional risk management to risk-taking.

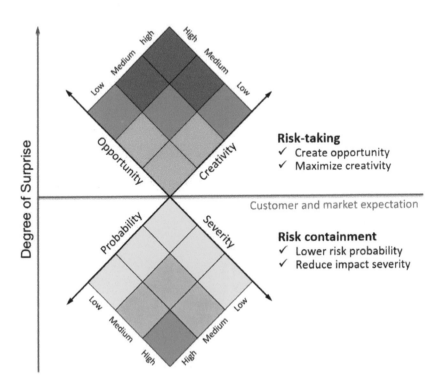

Chapter 7: Scheduling for Better Time Management

Theme: Time

<u>T Section</u>

Developed from the Critical Chain method, a new time management tactic combined with Critical Chain scheduling and buffer tracking can accelerate project schedule beyond expectations.

<u>M Section</u>

Eliminating due dates in scheduling reduces bad behaviors in time management.

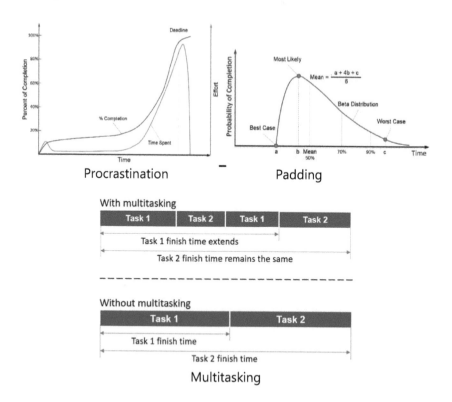

Procrastination — Padding

Multitasking

<u>L Section</u>

Build a project schedule using Critical Chain scheduling: no due dates and multitasking.

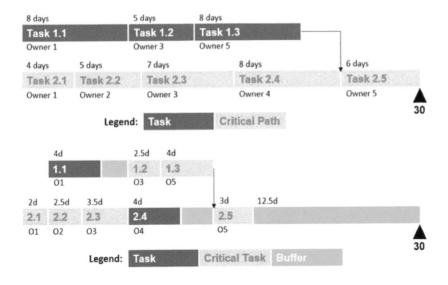

Tracking task and owner performance using buffer consumptions.

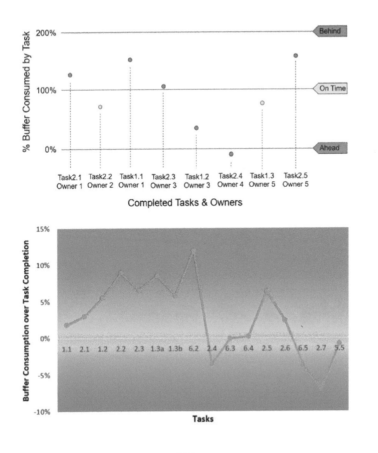

Chapter 8: Managing Path-finding Projects

Theme: Exploration

Apple the Six-P Model for managing Path-finding projects.

Understand the technology development cycle helps us to pursue PF projects with patience and persistence.

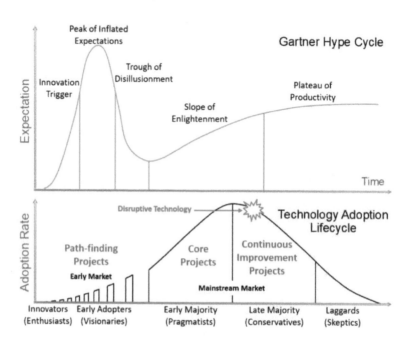

L Section

PF projects can be further categorized for better management.

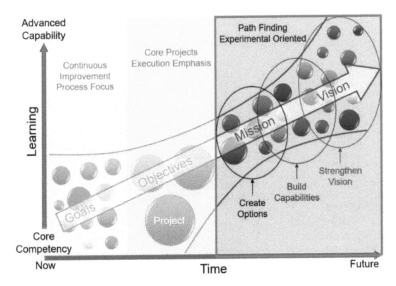

Checklist for Managing PF Projects
☐ Develop a vision (from passion)
☐ Define a mission (purpose statement)
☐ Identify targeted areas for learning
☐ Understand key technologies and their current lifecycle stages
☐ Initiate and map Path-finding projects based on purposes
☐ Strengthen vision
☐ Build capabilities
☐ Create options
☐ Develop high-level managing tactics
☐ Success measures
☐ External exposure
☐ Internal visibility
☐ Key resources
☐ Setup review committees
☐ Core project teams
☐ Key members, e.g. technical experts, executives
☐ Schedule review and grading cadence
☐ Conduct project review and grading sessions
☐ Performance grading
☐ Core project adoption potential
☐ Patent application and publication possibility
☐ Learning and sharing forums knowledge proliferation

Chapter 9: Managing Core Projects

Theme: Execution

<u>T Section</u>

Core projects must be carefully selected and receive the highest priority and focus.

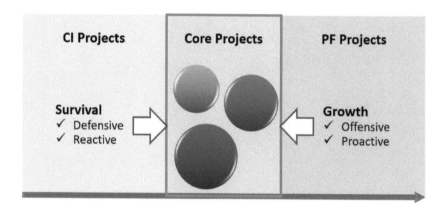

<u>M Section</u>

Core projects must be executed with quality and velocity.

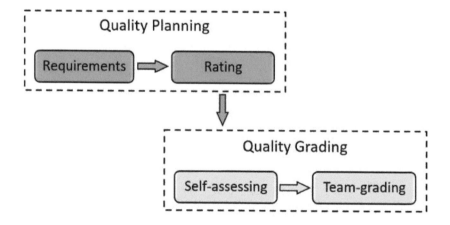

L Section

Utilize the Project War Room and modified Agile approach to ensure
Core project success.

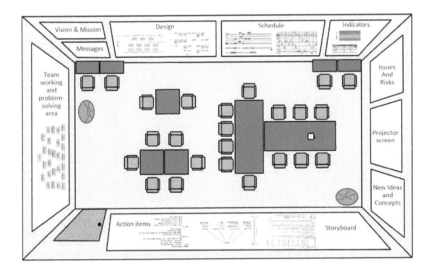

Checklist for Managing Core Projects
☐ Define the Core project strategy
☐ Survival, growth, or both
☐ Select Core projects
☐ Set priority
☐ Communicate to the entire organization
☐ Form Core project teams
☐ Core project team members
☐ Core project leaders
☐ Move the team from forming to performing quickly
☐ Logistics and ground rules
☐ Tasks and schedules
☐ Quality assessment and grading measures
☐ Practices for accelerating Core project velocity
☐ Establish Project War Rooms
☐ Design and layout
☐ Setup
☐ Rules and operating practices
☐ Practice Agile with simplicity and flexibility
☐ Roles and responsibilities
☐ Cadence
☐ Standardized activities

Chapter 10: Managing Continuous Improvement Projects

Theme: Efficiency

<u>T Section</u>

Continuous Improvement projects are imperative for setting up a culture of innovation. Drive CI project participation and engagement with three strategic angles: Lean, improvement challenges and career development.

<u>M Section</u>

Create Innovation matrix to drive projects in all business areas.

Innovation by Type / Innovation by Area	Radical Innovation	Reapplied Innovation	Incremental Innovation
Strategic Innovation		Project A	Project B
Business Model Innovation		Project C	Project D
Organizational Innovation	Project E	Project F, G	Project H, I
Marketing Innovation		Project J	
Product Innovation	Project K	Project L, M	Project N, O
Service Innovation		Project P, Q	Project R
Process Innovation	Project S	Project T	Project U, V, W
Supply Chain Innovation		Project X	Project Y, Z

Prioritize CI projects based on waste elimination and value creation.

Projects	Waste Elimination			Value Creation			Total	
	Probability	Severity	Score	Opportunity	Creativity	Score	Score	Priority
Project S			0.00	2	4	4.24	4.24	4
Project T			0.00	5	3	5.66	5.66	2
Project U	3	3	4.24			0.00	4.26	3
Project V	4	4	5.66			0.00	5.69	1
Project W	1	3	2.83			0.00	2.84	5

Scale: 0-6 with 6 the highest

L Section

Develop simple templates and processes to enable easy CI project initiation and execution, creating an autonomous culture where everyone participates in projects.

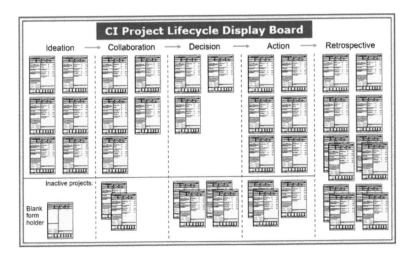

Checklist for Managing CI Projects

- ❑ Define participation and engagement strategy
  - ❑ Lean
  - ❑ Efficiency improvement
  - ❑ Career development
- ❑ Develop processes and templates
  - ❑ Innovation Matrix
  - ❑ Prioritization Worksheet
  - ❑ Project Initiation Form
  - ❑ Project Lifecycle Display
- ❑ Train employees
  - ❑ Innovation
  - ❑ Lean
  - ❑ Career development
  - ❑ Processes and templates
  - ❑ Tools for information, task and schedule management
- ❑ Apply CI project execution tactics
  - ❑ Visual displays
  - ❑ Daily reflections
  - ❑ Integration events
  - ❑ Mentoring and coaching
  - ❑ Reward and recognition

Chapter 11

Epilogue

Practicing the Concepts of This Book

It is now your turn to put the concepts into practice and this section is for you to write. Just keep in mind that there is no perfect tool for every situation. Experiment, modify and improve as you practice the concepts in your situation. Keep trying and surprise us: innovate.

I am looking forward to hearing your success stories as well as lessons from failures.

Chapter 11

Epilogue

Questions

It is also your turn to ask questions. Share them with me by E-mail
MBPM.Innovation@gmail.com.

There is also a web site for this book:
http://mbpminnovation.wordpress.com

Bibliography

1. Boston Consulting Group (2017). *The Most Innovative Companies 2016 Report.* URL: http://image-src.bcg.com/Images/BCG-The-Most-Innovative-Companies-2016-Jan-2017_tcm9-163125.pdf

2. Fast Company (2017). *Announcing the 2017 World's 50 Most Innovative Companies.* URL: https://www.fastcompany.com/3067756/announcing-the-2017-worlds-50-most-innovative-companies

3. Forbes (2017). *The World's Most Innovative Companies.* URL: https://www.forbes.com/innovative-companies/list/

4. IFI Claims (2017). *The World's Top Patent Producing Companies.* URL: https://www.ificlaims.com/rankings.htm

5. PM PrepCast (2017). *6th Edition Guide Release Date and Timeline.* URL: https://www.project-management-prepcast.com/free/pmp-exam/articles/853-pmbok-6-release-date-and-timeline

6. Greg A. Stevens & James Burley (1997). *3,000 Raw Ideas = 1 Commercial Success!* Research-Technology Management Journal, Vol 40, Issue 3. Taylor & Francis

7. Robert K. Wysocki (2013). *Effective Project Management: Traditional, Agile, Extreme,* 7th Edition. Wiley

8. Scaled Agile (2017). *SAFe Principles.* URL: http://www.scaledagileframework.com/safe-lean-agile-principles/

9. Live Science (2010). Personality Set for Life By 1st Grade, Study Suggests. URL: https://www.livescience.com/8432-personality-set-life-1st-grade-study-suggests.html

10. Daniel H. Pink (2011). *Drive.* Canongate Books Ltd.

11. Marcus Buckingham & Donald O. Clifton (2001). *Now, Discover Your Strengths.* The Free Press

12. David Epstein (2014). *Are the Athletes Really Getting Faster, Better, Stronger.* TED. https://www.ted.com/talks/david_epstein_are_athletes_really_gett ing_faster_better_stronger#t-53545

13. Aberdeen Group (2006). *Onboarding Benchmarking Report.*

14. Stephanie Vozza (2015). *Why You Should Never Paint Office Walls White.* Fast Company. URL: https://www.fastcompany.com/3044601/why-you-should-never-paint-office-walls-white

15. Kathleen D. Vohs (2013). *It's Not 'Mess.' It's Creativity.* The New York Times

16. Kathleen D. Vohs (2013). *Tidy Desk or Messy Desk? Each Has Its Benefits.* Psychological Science, Association for Psychological Science

17. Anna Steidle & Lioba Werth (2013). *Freedom from Constraints: Darkness and Dim Illumination Promote Creativity.* Journal of Environmental Psychology. Elsevier

18. Edgar H. Schein (1985). *Organizational Culture and Leadership.* Jossey-Bass

19. Josh Bersin (2015). *Culture: Why It's the Hottest Topic in Business Today.* Forbes

20. Chris Argyris & Donald A. Schön (1996). *Organizational Learning II: Theory, Method and Practice. Reading.* Addison-Wesley

21. Jeff Boss (2014). *4 Ways to Embrace Adaptability.* Forbes

22. Teresa Amabile & Steven J. Kramer (2011). *The Power of Small Wins.* Harvard Business Review

23. Marcela Litcanua, Octavian Prosteana, Cosmin Orosa, & Alin Vasile Mnerieb (2015). *Brain-Writing Vs. Brainstorming, Case Study For Power Engineering Education.* Procedia, Social and Behavioral Sciences. Elsevier, ScienceDirect

24. Tony McCaffrey (2014). *BrainSwarming, Because Brainstorming Doesn't Work.* Harvard Business Review

25. Stephen Young (2016). *Micromessaging: Why Great Leadership is Beyond Words.* McGraw-Hill Education

26. Bain Analysis (2011). *RAPID: Bain's Tool to Clarify Decision Accountability.* Bain & Company. URL: http://www.bain.com/publications/articles/RAPID-tool-to-clarify-decision-accountability.aspx.

27. Marcia Blenko & Jenny Davis-Peccoud (2011). *Great Decisions – Not a Solo Performance.* Bain & Company. URL: http://www.bain.com/publications/articles/decision-insights-10-great-decisions-not-a-solo-performance.aspx.

28. Cindy Perman (2012). *Hate Meetings? Why Most Are Complete Failures.* CNBC. URL: https://www.cnbc.com/id/48898453

29. Andrea Lehr (2015, updated 2017). *Why We Hate Meetings So Much [Infographic].* Hubspot. URL: https://blog.hubspot.com/sales/why-we-hate-meetings-so-much

30. John Walston (2015). *Finally! The Truth About Why We Hate Meetings [Infographic].* ResourcefulManager. URL: https://www.resourcefulmanager.com/why-do-we-hate-meetings/

31. Andrew S. Grove (2015). *High Output Management.* Vintage

32. George T. Doran (1981). *There's a S.M.A.R.T. way to write managements's goals and objectives.* Management Review, AMA Forum

33. Cyril Northcote Parkinson (1955). *Parkinson's Law.* The Economist

34. Eliyahu M. Goldratt (1997). Critical Chain. The North River Press

35. John M. Nicholas & Herman Steyn (2017). *Project Management for Engineering, Business and Technology.* 5th Edition. Routledge.

36. Stanford University (2018). *A Vision for Stanford*. URL: http://ourvision.stanford.edu

37. Gartner (2017). *Gartner Hype Cycle*. Research Methodologies, Gartner. URL: https://www.gartner.com/technology/research/methodologies/hype-cycle.jsp

38. Michael Mullany (2016). *8 Lessons from 20 Years of Hype Cycles*. LinkedIn Pulse. URL: https://www.linkedin.com/pulse/8-lessons-from-20-years-hype-cycles-michael-mullany/

39. Joe M Bohlen & George M. Beal (1957). *The Diffusion Process*. Special Report No. 18. Agriculture Extension Service, Iowa State College.

40. Geoffrey A. Moore (2014). *Crossing the Chasm, Marketing and Selling Disruptive Products to Mainstream Customers*. 3rd Edition. HarperBusiness

41. Leonard Cassuto (2013). *Ph.D. Attrition: How Much Is Too Much?* The Chronicle of Higher Education. URL: https://www.chronicle.com/article/PhD-Attrition-How-Much-Is/140045

42. Deanna Marie Putney (2003). *SWAT Team Composition and Effectiveness*. PhD dissertation, University of Tennessee.

43. Bruce W. Tuckman (1965). *Developmental sequence in small groups*. Psychological Bulletin.

44. The Myers & Briggs Foundation (2018). *MBTI Basics*. URL: https://www.myersbriggs.org/my-mbti-personality-type/mbti-basics/

45. Your Life's Path (2018). *DiSC Classic 2.0*. URL: https://www.thediscpersonalitytest.com/

46. Patrick Lencioni (2012). *The Advantage: Why Organizational Health Trumps Everything Else in Business*. Jossey-Bass.

CPSIA information can be obtained
at www.ICGtesting.com
Printed in the USA
LVHW011307270319
611964LV00033B/81/P

9 781733 555708